Solvent Engineering

Yaocihuatl Medina-Gonzalez

CNRS Research Scientist
Laboratoire du Futur (LOF)
UMR 5258 CNRS - Université de Bordeaux - Syensqo
Pessac, France

CRC Press
Taylor & Francis Group
Boca Raton London New York

CRC Press is an imprint of the
Taylor & Francis Group, an **informa** business

A SCIENCE PUBLISHERS BOOK

Cover credit: Image created by Yaocihuatl Medina-Gonzalez using Blender 2.93.

First edition published 2025
by CRC Press
2385 NW Executive Center Drive, Suite 320, Boca Raton FL 33431

and by CRC Press
4 Park Square, Milton Park, Abingdon, Oxon, OX14 4RN

Library of Congress Cataloging-in-Publication Data (applied for)

ISBN: 978-0-367-18588-6 (hbk)
ISBN: 978-1-032-93207-1 (pbk)
ISBN: 978-0-429-19705-5 (ebk)

DOI: 10.1201/9780429197055

Typeset in Times New Roman
by Prime Publishing Services

Preface

This book aims to provide readers with a comprehensive understanding of Solvent Engineering and its implications in chemistry, materials science, and chemical engineering. Solvents are ubiquitous in daily life, and professionals in chemistry, physical chemistry, and chemical engineering encounter them in almost every process they study. Consequently, scientists have endeavored to modify and control solvent properties to achieve their goals, rather than relying solely on the inherent properties of solvents.

A common initial approach to controlling solvent properties involves intuitively preparing a mixture of solvents. While this method is simple and effective for many applications, the need for more sophisticated solutions has led to the development of new solvents and systems with unique properties. Examples include supercritical CO_2, switchable solvents, and supramolecular solvents.

In addition to the demand for solvents tailored to increasingly specific applications, there is a pressing need for environmentally friendly options. Ionic liquids, deep eutectic solvents, and bio-based biodegradable solvents have been proposed to address the issues posed by volatile solvents.

Solvent Engineering cannot be approached without considering thermodynamics and mass and heat transfer, as the properties of the solvent medium strongly influence these phenomena. The aim of this book is to provide an insight into the breadth of Solvent Engineering, which has implications at both micro- and macro-scales. By understanding this field thoroughly, the judicious control and engineering of solvent properties can help overcome challenges in various fields, including biochemistry, chemistry, materials science, and more.

This book also gives a preliminary insight about the use of AI in the field of Solvent Engineering. Artificial Intelligence can significantly enhance Solvent Engineering by facilitating the collection of high-quality data and dramatically reducing the time needed to design solvent media with the required properties.

Acknowledgements

I would like to express my gratitude to all the funding institutions, in particular the Centre national de la Recherche Scientifique (CNRS) in France, that have supported the work on Solvent Engineering conducted by my group. I am also deeply thankful to the Masters' and Ph.D. students, as well as the postdoctoral researchers who have worked under my direction.

I extend a warm thought to my parents, who instilled in me a sense of responsibility, discipline, and a love for science. I am profoundly grateful to my husband, whose unwavering support has been instrumental throughout my career. Lastly, I hold a profound wish for my children: that my efforts have helped to shape a better world for you.

Contents

CHAPTER 1
Solvent Engineering
The Definition

Solubility is defined as the ability of a substance to form a solution with another substance, where the solution is a single, homogeneous gas, liquid or solid phase that is a mixture wherein the components are homogeneously distributed throughout the mixture. In general, the less abundant compound is called the solute, while the more abundant one is called the solvent. The properties of a solution are necessarily different from those of the pure compounds, for instance, the elevation of the boiling temperature or the decrease of the freezing point after the addition of a solute to a solvent, are two examples of physico-chemical properties that are modified when a compound is solubilized in a solvent.

The term 'Solvent Engineering' emerges from the awareness that the choice of the solvent impacts the performance of a process performed in solution. In industry, pure substances are very rare, and the properties of mixtures, together with mixing and demixing phenomena, are of prime importance for science and technology. It is also clear that the environmental footprint of a process is closely linked to the solvent used, as well as its technical and economic achievements too. It is then evident that the optimization of the solvent is the key to optimize chemical or physical processes that occur within this solvent.

In this context, Solvent Engineering can be defined as the design, building and manipulation of the solvent properties in order to optimize the chemical or physical processes performed within this solvent.

Optimization of solvent properties has for long been applied in separation techniques such as liquid-phase chromatography, high-performance liquid chromatography and liquid-liquid chromatography. However, even though these applications are excellent examples of the extents of Solvent Engineering, the treatment of these techniques is out of the scope of this book because of their vastness. Interested readers are referred to key references such as

(Fanali et al. 2017; Gilbert, Mary T. 1987; Palamareva 2005; Meyer 2010; Wingren and Hansson 2000) among others, for an insight on chromatography. The focus of this book will be on solvent engineering from the chemical and chemical engineering points of view and not on analytical applications concerning liquid-liquid separation. Even though chemical engineering is strongly concerned by this activity, several major reviews and books have been published before about this topic and the interested reader is strongly advised to consult them (Freedman 2020; Hyman et al. 2014; Stewart and Arnold 2008; Thornton 1992; Zhang and Hu 2013).

Before talking about the optimization of solvent properties, a recall of the properties of a solvent and of its characterization is needed. The following paragraphs will try to give some idea of the techniques used and of the different parameters employed to compare solvent power. However, this is not an exhaustive study of solubility phenomena and interested readers are encouraged to refer to some excellent documents already published in this field (Augustijns and Brewster 2007; Saal and Nair 2020; Wilhelm et al. 2007; Williams 2000) among others.

1.1 Characterization of Solvency Properties: Solvent Scales and Descriptors and their Relevance in Solvent Engineering

The characterization of solvents is important to rationalize their solvency properties and the solute-solvent interactions. For instance, the influence of solvents on reaction rates and in the band positions of the molecular spectra are undeniable effects arising from solvent physico-chemical and chemical properties. Activities such as separation and analysis through chromatography, chemical reactions, separation of products, and recrystallization, among others, need the understanding, knowledge and eventually prediction and control of solvent properties.

In a solution, there exists a competition between solute-solute, solvent-solvent and solute-solvent interactions, and hydrogen bonding abilities are the key in these interactions. There have thus been proposed two parameters: the hydrogen-bond donor (α) and hydrogen-bond acceptor (β) parameters for the solute, that can be established for the solvent too (α_S, β_S respectively). For a given solvent-solute system, interactions in solution are represented in Figure 1, defined by the hydrogen-bond donor and acceptor properties of the solvent and of the solute. In this figure, hydrogen-bond interactions are partitioned into four quadrants, with two quadrants representing solute-solute or solvent-solvent interactions dominating and where solubility is diminished.

Characterization of solvents is thus a necessary activity that has been developed in recent decades. Different techniques have been developed in

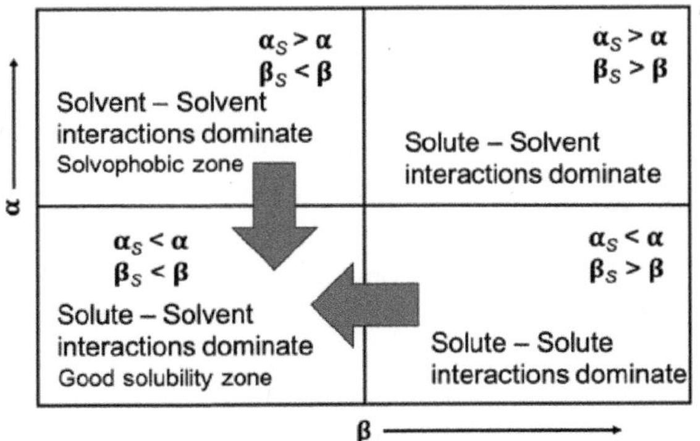

Figure 1. Generalized profile for hydrogen-bond interactions between a solute and a solvent in solution.

order to characterize relevant solvent properties such as partition coefficients, polarity, polarizability, acidity, basicity and permittivity, among others.

Several models and theories have been developed to elucidate the properties of solvents, and to help to predict solvent behavior. Among the strategies used, a number of solvent scales have been established based on physico-chemical observations and in molecular descriptors such as solvato-chromic effects, reaction enthalpies, reaction rates, and so on (Katritzky et al. 1999). Different solvent scales have been developed during the last 50 years with the aim of correlating solvent properties with solubility macroscopic phenomena observed. There are single parameter scales such as the Z scale of Kosower (Kosower 1958; Kosower and Ramsey 1959), $E_T(30)$ scale of Dimroth and Reichardt (Reichardt 1994), and S' scale of Drago (Drago 1992). Multi-parameter scales include the π^*, α, β scale of Kamlet and Taft (Kamlet et al. 1983), and the SPP, SA and SB scale of Catalán (Catalán 2009). Kosower was the first to propose a comprehensive scale used as a solvent polarity indicator. The Z polarity parameter was then defined as the molar transition energy for the charge transfer absorption band of 1-ethyl-4-(methoxycarbonyl) pyridinium iodide in a solvent. The $E_T(30)$ scale of Dimroth and Reichardt uses 2,6-diphenyl-4-(2,4,6-triphenyl-1-pyrido) phenoxide as the standard probe (Reichardt 1994); this molecule exhibits a strong solvato-chromic character. The $E_T(30)$ parameter is defined as the position of the maximum of the first absorption intra molecular charge transfer band of this molecule in the solvent. Reichardt's dye, as this molecule has been called, presents a strong hypsochromism when the solvent polarity increases, as its dipole moment in the excited state is much smaller than in the ground state. The parameter E_T^N is obtained by normalizing the $E_T(30)$ scale by assigning a zero value to tetramethylsilane (TMS) and a unity value to water, as seen in Equation 1.

$$E_T^N = \frac{E_T(Solvent) - E_T(TMS)}{E_T(Water) - E_T(TMS)} = \frac{E_T(Solvent) - 30.7}{32.4}$$ Equation 1

This scale is the most used, with values for more than 300 solvents already determined. Kamlet et al. developed a multi-parameters scale based on UV-Vis spectral data of solvato-chromic probe molecules, in which β, a scale of solvent hydrogen-bond acceptor (HBA) basicity; α, a scale of solvent hydrogen-bond donor (HBD) acidity and π^*, a scale of solvent dipolarity/polarizability are evaluated. Here, π^* measures the ability of the solvent to stabilize a charge or a dipole by effect of its electric dielectric effect, β refers to the ability to accept a proton in a hydrogen bond, and α refers to the ability to donate a proton in a hydrogen bond (Kamlet et al. 1979a; 1979b; 1977c; Kamlet and Taft 1976; Marcus et al. 1988; Taft et al. 1982; Taft and Kamlet 1976; Taft et al. 1985; Yokoyama et al. 1976). These and other solvent scales have been thoroughly reviewed and discussed in literature. Notably, more than 180 solvent scales have been identified in literature (Katritzky et al. 2004). One of the challenges is to organize, and, eventually, to rationalize all these scales in order to apply them in the most judicious manner to understand and predict solvent effects. In particular, an interesting feature of solvent scales is that considerable efforts have been directed recently in trying to relate complex solvent effects to fundamental molecular properties or descriptors. Descriptors have been largely used as a means to relate the chemical structure with specific types of inter-molecular interaction. There exist thermodynamic, quantum chemical, topological, electrostatic and charged partial surface area descriptors, among other types (Katritzky et al. 2005). Quantum chemical descriptors are widely used in solvent scales. They can be classified as:

- Descriptors related to cavity formation
- Descriptors related to electrostatic interactions
- Descriptors related to dispersion interactions
- Descriptors related to specific interactions (hydrogen bonding)

According to A. Katritzky (Katritzky et al. 2005), and as an example, the descriptors related to electrostatic interactions in liquid media have the largest occurrence when studied through QSPR (quantitative structure – property relationship) methodology. This methodology and its relationship with Solvent Engineering will be addressed in a following section.

All these solvent scales and descriptors have been established though experimental methods or though theoretical calculations; modeling of solvents and of solvent properties.

From a Solvent Engineering point of view, usage of solvent scales may represent a strategy to control the properties of the solvent media to improve a process carried out in the considered solvent or mixture of solvents. Some examples can be given here without the intention to be exhaustive, as more details in applications of Solvent Engineering will be treated in the following sections. Self-assembly of gel-phase materials is strongly dependent on the solvent environment. Hirst and Smith (Hirst and Smith 2004) observed that the solvent environment plays a key role in modulating the hierarchical self-assembly of dendritic building blocks into a gel-phase material when studying the self-assembly of diaminododecane with dendritic L-lysine-based peptides in different solvents. The solvent induced subtle changes in the mesoscale aggregate morphology and macroscopic which induced different behavior of the self-assembled state, in particular concerning the gel-sol transition temperature (T_{gel}). Also, circular dichroism spectroscopy revealed that helical self-assemblies were present even in solvent systems that do not form gel phase materials, which was interpreted as a capacity of the solvent environment to modulate the level of hierarchical self-assembly. The rationalization of the effect of the solvent was performed through the use of the Kamlet-Taft multi-parameter scale, where the parameter related to the capacity of the solvent to donate hydrogen bonds (α) was found key in the capacity of the solvent to disrupt the gel formation, while the other parameter in relationship with the polarity of the solvent (δ_a) showed the effect of the polarity of the solvent in the observed changes in T_g (Figure 2). This work represents an effort towards a rationalization of the solvent effects through solvent scales in order to control the macroscopic properties of complex systems such as gels.

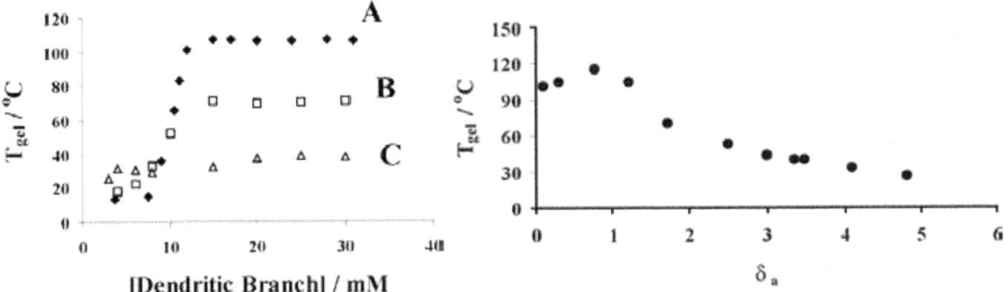

Figure 2. (a) Effect of the solvent environment on the gel-sol transition temperature (T_{gel}): (A) toluene, (B) 1,2,3,4-tetrahydronaphthalene, (C) 1,2-dichlorobenzene. (b) Effect of the polarity solubility parameter of the solvent (δ_a) on T_{gel}. Reprinted with permission from Hirst, A. R. and D. K. Smith, Langmuir 2004, 20: 10851–10857, Copyright (2004) American Chemical Society.

Other examples of the possibilities of using solvent scales for the optimization of solvency and transport properties though a QSPR approach are exemplified by the good amount of works concerning solvent effects on reactivity. For instance, the work of L. Moreira et al. (Moreira et al. 2019) has shown that a QSPR approach, established through multiple linear regressions based on the Kamlet-Taft solvent scale can provide predictive quantifications of the interactions between solutes and solvent prevailing in the heterolysis of tertiary alkyl halides, by using a set of 21 protic and aprotic solvents. It is notable that the application in this study relates the three solvatochromic Kamlet-Taft parameters (α, β and π^*) to the reaction rate constant (k) through an equation of the form shown in Equation 2:

$$\log k = a_0 + a_1\,\pi^* + a_2\,\alpha + a_3\,\beta \qquad\qquad \text{Equation 2}$$

with a_i being coefficients obtained through multiple linear regression. For comparison, another equation was proposed in this study by correlating π^* and α descriptors with the Dimroth-Reichardt E_T^N scale as well as the Katritzky model, which considers the E_T^N value and a descriptor $f(\varepsilon)$ as a measure of the solvent polarity (Equation 3).

$$\log k = a_0 + a_1\,E_T^N + a_2\,f(\varepsilon) \qquad\qquad \text{Equation 3}$$

This study shows that controlling the properties of the solvent may allow a control on the reaction rate of some chemical reactions, which can be rationalized by using different descriptors and solvent scales through a mathematical equation. Other examples of this approach will be presented in subsequent sections.

S. Spange and N. Weiß (Spange and Weiß 2023) analyzed the behavior of UV-Vis solvato-chromic probes for the hydrogen bond donor property in alcoholic solvents by introducing the hydroxyl group density (D_{HBD}) parameter instead of the acidity parameter pKa. The D_{HBD} takes into consideration the concentration of the OH groups per volume of the solvent without taking into consideration the hydrogen bond donating character in terms of acidity. The authors compare the ability of different solvent scales to really reflect the acidity of the solvent and not the sum of the interactions of the OH dipoles or the global polarity of the hydrogen bond network in terms of electrostatic forces. Among their conclusions, the authors declare that the original definition of the HBD property of alcohol as has been proposed by Kamlet-Taft, is not accurate, as it does not relate to acidity properties of the solvent, and that the adjustment proposed by Guntmann could be used to solve this problem.

1.2 Modeling of Solvents and Solvent Engineering

The study of solutions and their properties, together with the prediction of solvent effects on chemical reactions on UV absorbance spectra of probe molecules, to cite some macroscopic phenomena, have been major concerns during the last century. Modeling of solvents and of solutions has been performed through different approaches including empirical observations, thermodynamics, solvent scales, and descriptors, among others. For industry, accurate modeling of solutions and the phenomena involved in solution process is the key to design, optimize and operate process units and even entire chemical processes in fields as varied as oil, gas, petrochemicals, pharmaceuticals and agriculture, among others.

Modeling of solvent properties can be performed by applying different models, depending on the degree of accuracy and complexity needed, together with the calculation difficulty desired: ideal, regular or real solutions models. This can be addressed through thermodynamic approaches that include equations of state such as Peng-Robinson, and Redlich-Kwong, as well as equations of state based on the statistical associating fluid theory (SAFT). Other methods for solvent properties modeling rely on the use of solvent scales and descriptors, some of which have been presented in the following sections. As this section is not intended to be exhaustive by presenting all the existing models able to describe or predict solubility processes or solvent effects, more adapted literature can be found presenting good and accurate descriptions of each of these methods. This section is mostly intended to connect modeling of solvents with Solvent Engineering. For instance, the strategy presented in the work of M. Veit et al. (Veit et al. 2019) concerning the development of equations of state from first principles calculations through the use of machine learning for the methane molecule can be extrapolated to more complex systems. In this study, the total interaction potential was separated into different components concerning repulsion, dispersion, electrostatic and induction terms. Quantum calculations and a Gaussian Approximation Potential (GAP) machine learning method were used to obtain the potential energy surface (PES), which was fitted by a Gaussian process regression. This methodology allowed more efficient simulations, making it possible to perform large, expensive liquid simulations, even with quantum nuclear effects, at the level of many-body dispersion-corrected DFT, which was impossible with other methodologies. The methodology used in this study can be transposed to more complex systems to perform calculations of solubility by using mixtures of solvents with the aim to obtain specific properties able to give the desired solvency or transport properties. In this optic, *in silico* design of solvents with the desired properties of toxicology, solvency and other properties can be envisaged, as has been exemplified in the work presented by

L. Moity et al. (Moity et al. 2016), concerning the design of glycerol-based solvents for nitrocellulose through a procedure combining Computer Assisted Organic Synthesis (CAOS) and Computer Aided Molecular Design (CAMD). The top-down approach used in this study consists of

1) defining the properties that the solvent should comply with, in this case, the solubilization of nitrocellulose;

2) generating *in silico* the structures that best fit the desired specifications.

The second step of this approach was performed through the utilization of a software called GRASS (Generator of Agro-based Sustainable Solvents) (Moity et al. 2014), which is a tool to allow a series of chemical transformations to valorize bio-based building blocks into target intermediate or commodity chemicals such as solvents. The GRASS software works as a virtual reactor, where a molecule, in this case, glycerol, allows the obtaining of plausible generic structures. The obtained molecules are used as inputs for the third step of the approach, where the software IBSS (InBioSynSolv was used). This CAMD software uses an evolutionary genetic algorithm where molecular structure and composition are used together as decision variables to find binary aqueous mixtures with target solvency properties (Heintza et al. 2012). It must be noted that the ability of the *in silico* designed solvent to solubilize nitrocellulose was evaluated through the Hansen approach; which can, in our opinion, be improved by using more accurate descriptors.

Solvent Engineering concerns rational solvent selection and eventually, the modification of the solvent environment to optimize the targeted process. In the work of Y. Amar (Amar et al. 2019), the selection of a solvent in a rational fashion was exemplified by the application for the catalyzed asymmetric hydrogenation of a chiral α-β unsaturated γ-lactam, where two simultaneous objectives were designed: high conversion and high diastereomeric excess (d.e.). The authors of this study propose that solvent descriptors can be incorporated into the self-optimization algorithm of the reaction to create predictive models. The method used is illustrated in Figure 3. In this strategy, physical knowledge about solvents is introduced to the machine learning method through a set of molecular descriptors that have their origin in literature, or from computational sources such as COSMO*therm*, which combines quantum chemistry and thermodynamics to predict properties. Some molecular descriptors used in this study were viscosity, σ-profiles and octanol-water partition coefficient, among others. The machine learning algorithm used the Thompson sampling efficient multi objective optimization (TS-EMO) algorithm developed by E. Bradford et al. (Bradford et al. 2018). In this study, sigma profiles were segmented in 3 or in 5 segments, and two models were developed in each case. In the case of

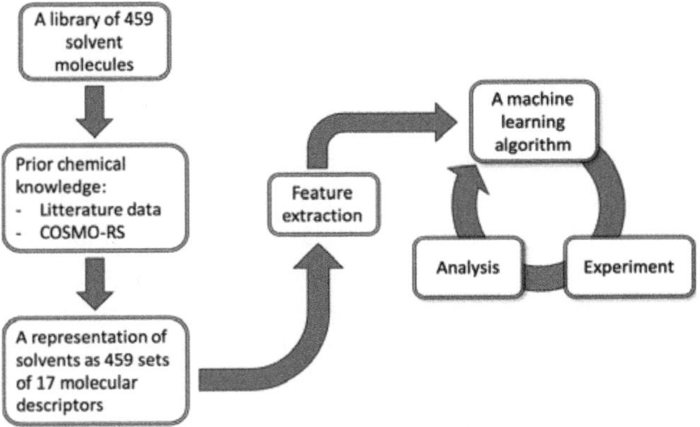

Figure 3. The workflow and transition of chemistry knowledge into machine learning domain and process domain, i.e., experiment and analysis. Adapted from Y. Amar et al. (Amar et al. 2019).

conversion, the model using σ-profiles segmented in three was more accurate, while for diastereomeric excess, prediction appears to be better using the model with the σ-profiles segmented in five. This can be explained by the fact that solvent polarity, which is information contained in the profile extremes, has more impact in d.e. and is better described by σ-profiles segmented in five rather than in three (Amar et al. 2019).

The approaches presented are just few examples of *in silico* design methods that can be, in the end, robotized and automatized to obtain *ad hoc* solvents with solvency, transport and toxicological characteristics desired for a target application.

A great quantity of data has been generated concerning molecular properties, and in recent years, QSPR (Quantity Structure – Property Relationship) approaches have emerged to predict solvent effects or to select solvents in a rational way. For instance, T. N. Borhani et al. (Borhani et al. 2019) have proposed a methodology for the construction of QSPR models that can be used to predict the free energy of solvation for a solute/solvent pair. The proposed methodology is a hybrid between theory-based and experiment-based QSPR methods. Experimental descriptors or properties were used for the description of the solvent, such as, the boiling point, the relative permittivity, the surface tension, and the octanol-water partition coefficient; while quantum mechanics descriptors were used for the selected solutes, such as electronic basicity, acidity, energy of the HOMO, of the LUMO, and dipole moment, among others. Interestingly, polarizability, molecular van der Waals volume and entropy of the solute are the most important properties for the model, in order of decreasing relevance. Concerning the solvent properties, the dipole moment, the octanol-water partition coefficient and the surface tension appeared to have the greatest weight in the model. On the other hand, the heat of vaporization and molar volume of the solvent appeared to have limited importance, which is surprising as they are frequently-used properties in LSER models and in widely used models such as the Hildebrand solubility parameter.

With the aim to improve the performances of bulk hetero junction solar cells fabricated from $[C_{60}]$ and $[C_{70}]$ fullerenes and from their derivatives, P. A. Troshin et al. (Troshin et al. 2009) have studied the relationship between the material solubility and the performances of the cells. In particular, they have studied the solubility of a number of fullerenes derivatives in different solvents and the performance of their blends with the polymer P3HT (Poly(3-hexylthiophene-2,5-diyl)) in photovoltaic devices. Interestingly, solubility of the fullerene derivatives was not only found dependent on the appended groups but also on the solvent used for the synthesis of each compound, which can be entrapped into the cavity of the fullerene, forming an adduct and bestowing particular solubility properties and giving differences in performances of the fabricated devices. Among the optimization parameters used for the fabrication of the devices, annealing time and temperature as well as fullerene: polymer ratio and spin coating frequency were studied. According to the authors, polymer phase and crystalline nano domains formation can be obtained by using different solvents and can change the performances of the device. Fullerene solubility was directly related to solar cell output parameters (short circuit current (I_{sc}), open circuit voltage (V_{OC}), fill factor (FF), and power conversion efficiency (η)). All the properties increase with fullerene solubility up to a maximum and then decrease or stagnate. As for the surface state, after annealing of the mixtures, some roughening was provoked by the annealing. However, an increase in solubility leads to a less rough surface; inversely, poor solubility brings the formation of fullerene aggregates and phase separation of the polymer-fullerene system, impacting performance. However, a minimum of phase separation was needed to obtain good performances, as observed with the mixture with chlorobenzene, which is an excellent solvent for fullerenes but gave poor performances when used in the devices.

A. P. Toropova et al. (Toropova et al. 2011; 2019) have proposed QSPR models for the solubility of $[C_{60}]$ and $[C_{70}]$ fullerenes derivatives in chlorobenzene. For that, a software called CORAL (Correlations and Logic) has been developed, which is based on the representations of the molecular structures called SMILES (simplified molecular input line entry systems) and on QSPR approaches and a Monte Carlo optimization algorithm. This study was motivated by the importance that fullerene and fullerene derivatives have in industry. Results obtained were in good accordance with experimental data.

H. Lim and Y. J. Jung (Lim and Jung 2019) have developed a deep learning model for solvation-free energies in generic organic solvents (DELFOS).

Delfos was presented as a machine learning-based QSPR method. Here, two mathematical functions are used:

i)   encoding the chemical structure into a molecular descriptor and

ii) mapping to predict the property or activity.

Concerning the encoding function, various molecular descriptors have been proposed to represent a molecule. For instance, empirical properties such as molecular weight and number or hydrogen bond donor and acceptor sites; molecular fingerprints such as SMILES and graphical representation. Concerning mapping functions, properties were extracted from the encoded molecular features. The authors combined QSPR approaches with a recurrent neural network model to calculate solvation energies. Solvation energies of 2495 pairs of 418 solutes and 91 solvents were calculated and the results compared with theoretical approaches such as molecular dynamics with excellent accuracy when the neural network was trained with sufficiently varied chemical structures.

CHAPTER 2
Strategies used in Solvent Engineering

Optimization of solvent properties can be performed by several strategies that can use physical, chemical or both approaches. From the simplest strategy that is the mixture of two or more solvents to more sophisticated approaches such as the use of a chemical reaction to modify the used solvent, these strategies can be used alone or together to obtain the needed properties in a solvent. These strategies have been applied to different fields, opening doors to new fascinating applications. In the following chapters, some possible strategies for Solvent Engineering will be treated, some of which have already been cited in literature; as well as the different fields where Solvent Engineering has started to appear and be used. After that, some Perspectives will be given in order to present our vision about the future of Solvent Engineering.

2.1 Supercritical Fluids (SCFs)

A fluid is in supercritical state when the temperature and pressure is higher than its critical-point values (Tc and Pc respectively). In this state, the fluid presents one single phase, with a density appropriate to dissolve other compounds and a diffusivity of solutes higher to that in liquids. Also, the viscosity of SCFs is lower, which increases mass transport. Near the critical point, SCFs present high compressibility, which involves important changes in density and, as consequence, in solubility properties. The density of SCFs strongly influences the solubility of a solute; some studies have shown that solubility depends exponentially on density (Eckert et al. 1996), which means that a small variation in pressure near the critical point may provoke great solubility changes. This property allows us to envisage Solvent Engineering by using SCFs and was one of the first examples where the interest of a solvent with adjustable and reversible properties was evidenced.

It is interesting to treat some of the aspects of the thermodynamics of supercritical fluids, in order to understand the implications of the behavior of these fluids in the applications already developed, and to imagine new fields where these properties can represent a breakthrough, in, for instance, materials science, chemical reactions and extractions, to cite a few.

Three characteristics define the critical point:

1) the difference between gas and liquid disappears,
2) the compressibility diverges and
3) the critical opalescence appears.

Indeed, when a liquid substance in equilibrium with its vapor phase is heated above its critical temperature (Tc) and pressurized above its critical pressure (Pc), the interface between the gas and the liquid disappears and one phase remains: a supercritical fluid. This fluid presents specific physico-chemical properties such as viscosity close to a gas and density similar to a liquid, which represent advantages in the chemical process by increasing mass transfer.

Concerning compressibility, Equation 4 expresses the conditions for the divergence of the compressibility

$$\left(\frac{\partial P}{\partial V}\right)_T = 0 \quad \text{and} \quad \left(\frac{\partial^2 P}{\partial V^2}\right)_T = 0 \qquad \text{Equation 4}$$

The isothermal compressibility $K_T = \left(\frac{1}{\rho}\right)\left(\frac{\partial \rho}{\partial P}\right)_T$, that is the change in volume as pressure changes at isothermal conditions, also diverges at the critical point. Indeed, K_T can be related to $\int r^2 [g(r) - 1]dr$, where $g(r)$ is the pair correlation function, which is the ratio of the local to bulk density at a distance r from a molecule fixed at the origin. At the critical point, $g(r)$ becomes long-ranged, beyond the extent of the range of molecular potentials. The long-range correlations induce higher K_T variations, influencing the rate of change of solubility with respect to temperature and pressure. Compressibility of SCFs needs to be taken into consideration when studying cavitation in these fluids. This is because it has been observed that cavitation intensity (defined as the bubble ratio of maximum radius before collapse to minimum radius after collapse), decreases to more than half when taking into consideration compressibility of the supercritical CO_2 and compared to calculations without including the compressibility of the medium (Chen and Lu 2015).

Density fluctuations are described by ξ (Tucker 1999), who calls it the correlation length. The pair correlation function near the critical point (but not in the critical point) is given by:

$$g(r) \propto \frac{\exp\left(\dfrac{-r}{\zeta}\right)}{r} \qquad\qquad \text{Equation 5}$$

ξ diverges at the critical point for any fluid, exceeding molecular dimensions in the vicinity of the molecule. Thus, compressibility augmentation, correlation length increase and fluctuation-dominated long-range behavior of the correlation function are not related to short-ranged details of the intermolecular force which govern solvation. These short-ranged effects can be affected by using a co-solvent which can be appreciably different from the SCF used concerning its size, polarity and hydrogen bonding character among other properties. These differences allow a control of the local solvation properties that can be profitably used in separations, reactions, and so on, even if around the critical point, process control can be delicate.

Density and viscosity of SCFs are of crucial interest when talking about Solvent Engineering, and in general for chemical processes. Density has an important role in solubility phenomena, and viscosity in mass transfer. In this view, it is worth presenting some specificities of these properties in SCFs. Density and viscosity of SCFs has been studied for a long time. On one hand, equipment adapted to high pressures has been designed to obtain experimental data in a more and more accurate fashion; on the other hand, increasingly perfected calculation techniques and models have been used to predict the behavior of these properties as a function of temperature and pressure. These methods include empirical correlation, equations of state, and molecular modeling techniques, among others (Bahadori et al. 2009; Böttcher et al. 2012; Langenfeld et al. 1992; Nishikawa et al. 2004; Span and Wagner 1996; Tilly et al. 1994; Wang et al. 2015). For example, density fluctuations of SCFs such as CO_2, CHF_3 and H_2O have been measured by small-angle-X-ray scattering experiments (Nishikawa et al. 2004; Saitow et al. 2004) and compared with the results obtained from calculations from an equation of state (Span and Wagner 1996) and from molecular dynamics (Goodyear et al. 2000). These studies clearly demonstrate the inhomogeneous structure of supercritical fluids and, overall, the effect of clustering, which is of crucial importance when the energy between solutes and solvents is calculated for chemical reactions in SCFs and for computer simulations. This clustering effect has been predicted in water and in mixtures of water-CO_2 in supercritical conditions through molecular dynamics calculations (Kalinichev and Churakov 1999). These density fluctuations are at the origin of the high compressibility of GXLs. Density inhomogeneity of SCFs around a solute molecule has been evidenced through spectroscopic measurements (solvato-chromic shifts of probe molecules, IR shifts, FTIR absorption, fluorescence anisotropy to cite some) and modeling (Tucker 1999).

On the other hand, it has been recently shown that a supercritical fluid is not homogeneously distributed but that two distinct regions can be differentiated with gas-like and liquid-like properties. These two regions, that are reminiscent of the subcritical domains, are divided by an extension of the co-existence line called the Widom line (Banuti 2015; Fomin et al. 2015; Raju et al. 2017; Simeoni et al. 2010). It has been observed that impurities in pure SCFs can shift the location of the Widom line in CO_2 (Imre et al. 2015) and in water (Imre et al. 2011). In the context of Solvent Engineering, modulation of the position of the Widom line gives a control on the gas-like and liquid-like properties of a SCF. However, more research is needed to exploit this possibility, as work on the determination of Widom lines for pure supercritical fluids and their mixtures is at present limited.

With the generalization of high performance computers, one technique that has gained recent interest for the calculation of density, viscosity and other properties such as self-diffusion coefficients in SCFs is molecular dynamics (Belonoshko and Saxena 1991; Branam 2005; Duan et al. 1996; Galliéro et al. 2005; Tsuzuki et al. 1996; M.S. Zabaloy et al. and Macedo 2005). This technique is of interest to gain insight and, eventually, for the prediction of properties of supercritical fluids, given the technical difficulty that measurements at high pressure can represent, for instance for viscosity, self-diffusion coefficients, binary diffusion coefficients, compressibility and clustering effects, to cite some (Kalinichev and Churakov 1999; Sakuma and Ichiki 2016; Yoo et al. 2012).

Experimental measurements of the viscosity of SCFs other than CO_2 are scarce (Fields et al. 2011; Li and Xing 2014). However, some empirical correlations have been used to correlate the existent experimental data (Heidaryan et al. 2011; Tilly et al. 1994). Some theoretical models have been used as well to calculate the viscosity of multi-component fluid mixtures forming a SCF, such as the rigid-sphere model (Vesovic et al. 1998), sometimes by using the Lennard-Jones intermolecular potential (LJ, Equation 6) (Marcelo S. Zabaloy et al. 2005), where r is the intermolecular distance, u the intermolecular potential energy, σ the Lennard-Jones separation distance at zero energy, and ε the depth of the Lennard-Jones potential well.

$$u(r) = 4\varepsilon \left[\left(\frac{\sigma}{r} \right)^{12} - \left(\frac{\sigma}{r} \right)^{6} \right] \qquad \text{Equation 6}$$

The Lennard-Jones intermolecular potential is widely used as it is simple but realistic: the observed behavior of the viscosity for real fluids is qualitatively well reproduced by this potential over a wide range of conditions. Reduced properties (Temperature (T^*), Pressure (P^*), density (ρ^*) and viscosity (η^*)) can be defined as:

$$T^* = \frac{kT}{\varepsilon}, P^* = \frac{P\sigma^3}{\varepsilon}, \rho^* = \frac{N}{V}\sigma^3 = N_A\,\rho\sigma^3, \eta^* = \eta\frac{\sigma^2}{\sqrt{m\varepsilon}}$$

where k is the Boltzmann constant, T and P are the absolute temperature and pressure respectively, N is the number of molecules, V is the volume of the system, N_A is the Avogadro number, ρ is the density, η the sheer viscosity and m is the mass of one molecule. Figure 4 shows the density of a LJ fluid as function of temperature.

A comparison between the calculated viscosity of a LJ fluid and experimental data for methane as function of pressure at different temperatures can be found in the work by Zabaloy et al. (Zabaloy et al. 2001), while other examples too have been reported in literature (Ravi and Guruprasad 2008; Zabaloy et al. 2005). It is clear from this figure that even if the Lennard-Jones fluid is a fluid made of spherical molecules, it remains a good reference not only for qualitative studies but because a good amount of thermodynamic data obtained from molecular simulations has already been reported.

Thermal conductivity in SCF presents a large increase near the critical point as shown by B. Le Neindre et al. (Le Neindre et al. 1984) for different compounds including CO_2, ammonia and ethane, among others. These authors present measurements of thermal conductivity together with a method of calculation to obtain it with good accuracy. A method to calculate the thermal conductivity for mixtures is also proposed in the cited work.

Thermal conductivity of CO_2 as function of temperature at different pressures presents a divergence around the critical point; the critical enhancement of thermal conductivity of propane as a function of density at

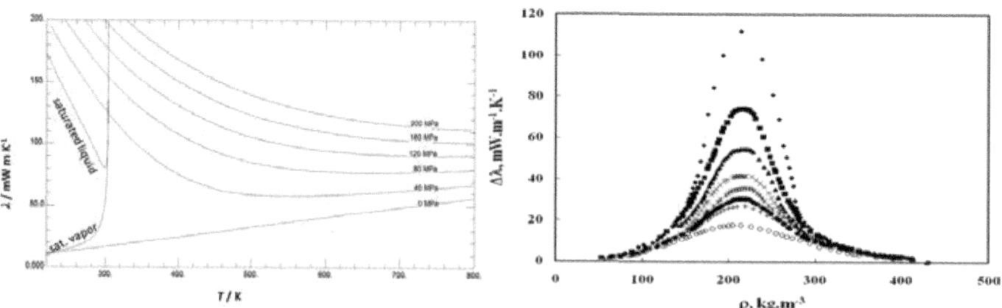

Figure 4. (a) Thermal conductivity of CO_2 as a function of temperature for different pressures and (b) Distribution of the critical enhancement of the thermal conductivity of propane as a function of density along eight isotherms at ◆ T = 370.45 K; ■ T = 371.05 K; ▲ T = 371.97 K; × T = 373.20 K; ∗ T = 374.28 K; ● T = 375.10 K; + T = 375.90 K; and ○ T = 381.00 K. Reprinted with permission from (a) M.L. Huber et al. (Huber et al. 2016) and (b) B. Le Neindre et al. (Le Neindre et al. 2014).

different temperatures. It is interesting to note that pressure and temperature allow the control the thermal conductivity of SCFs in a large range of values. Unfortunately, data on thermal conductivity of mixtures concerning SCFs is extremely scarce.

2.2 Supercritical Fluids (SCFs) Modified by a Co-solvent

As the solvency characteristics of CO_2, the most commonly used SCF, are not adapted for polar solutes because of its non-polarity, adding co-solvents to CO_2 to obtain a mixture that keeps some of the properties of supercritical CO_2 (low viscosity and high density) to increase its solvency power, is a strategy that has been often used. In a way, it allows the performance of Solvent Engineering, in order to adapt the properties of the solvent to the process happening therein.

The effect of pressure on solvency properties of SCFs modified by co-solvents has been thoroughly studied from a fundamental point of view (Curran et al. 2006; Knez et al. 2018; Lou et al. 1997; Yizhak Marcus 2006). Different applications such as the extraction of active compounds from vegetable matrices (Khaw et al. 2017; de Melo et al. 2014; Sovová Helena and Stateva Roumiana 2011) and in chemical reactions (Brunner 2004; Knez et al. 2018; Skouta 2009), among others, have also been deeply discussed in well-documented reviews, and interested readers are encouraged to consult them (Mohammad and Inamuddin 2012).

When a SCF is modified by a co-solvent, or, in other words, when a homogeneous mixture of a solvent in low concentration and a SCF is prepared, density, viscosity, and the other physico-chemical properties of the initial solvent are modified. In particular, dramatic enhancements in the solubility of a solute can be obtained by the addition of co-solvents to SCFs. This effect comes from the increase in density of the mixture and from the chemical interactions between solute and co-solvent (Solubility enhancement in supercritical solvents 1990), together with the mass transport increase brought by SCFs. Previous studies have shown that the increase in density, compared to pure SCFs, is more drastic in the zones of high compressibility, meaning low pressure, while in the zones of high pressure, density of the mixture is close to that of the pure SCF (Yun et al. 1996). In mixtures of SCF-co-solvent, density and concentration fluctuations are present. Density fluctuations arrive with the inhomogeneity of molecular distribution: important density fluctuations indicate strong aggregation. Concentration fluctuations happen with the inhomogeneous mixing of different molecules: large concentration fluctuations denote that identical molecules get together.

In the context of Solvent Engineering, and as co-solvents have been long used to enhance the solvation properties of SCFs, different research groups have studied the effect of several co-solvents on the solubilization of some solutes with the aim of rationalizing this effect. Some examples are the comparison of the solubility of some solutes in supercritical CO_2 and in supercritical CO_2-methanol (95/5% mol), and where enhancement of only one of the solutes has been observed in the mixture (Van Alsten et al. 1984). Later, it was concluded that solubility enhancement was not related to polarity of the co-solvent, but the increase in solubility was attributed to a complex formed between methanol and the solute (Walsh et al. 1987). Other examples of formation of complexes between co-solvents and solutes can be found in literature (Schmitt and Reid 1986). According to Walsh et al. (Walsh et al. 1987), the two carbonyl groups present in CO_2 molecule can compete with other hydrogen bond acceptors for the hydrogen bond donors present in the system. This behavior needs to be taken into account when a supercritical CO_2-co-solvent-solute system is studied or envisaged for a chemical process. On the other hand, if well understood and profitably used, this competence between interactions opens doors to fine control of processes at the micro scale that can allow applications in fascinating areas such as crystallization (Bertolasi et al. 2001), innovative chemical reactions (Martins et al. 2017), and even self-assembly (Yamamoto et al. 2018), where control of hydrogen bonding acceptor and donor characters are crucial to guide the process.

Other properties that can be controlled by adding a co-solvent to a SCF are viscosity, dielectric properties and thermal conductivity. The effect of a co-solvent on the viscosity of mixtures of supercritical CO_2 and co-solvent has been studied by different research groups and some results are shown in Figure 5. As can be observed, viscosity of mixtures of supercritical CO_2 and co-solvent can be controlled in an important game of values through temperature and pressure (Figure 5a). Besides these two variables, co-solvent molecules need to be carefully chosen in order to obtain the desired effect (Figure 5b), as a subtle change in molecular weight can have dramatic effects on the viscosity of the mixture.

Concerning dielectric properties, and in particular dielectric constants, co-solvents can strongly modify this property when added to a supercritical fluid and in particular CO_2, as observed in Figure 6. The concentration of co-solvent and the chemical structure are two variables, together with temperature and pressure, which can be used to control this property. It is to be noted that the addition of a polar compound such as ethanol imparts great changes in dielectric constants of the mixture as function of co-solvent volume fraction in contrast with the small changes in dielectric constant obtained by temperature and pressure adjustments in pure CO_2 (Figure 6a,b,c). As can be imagined intuitively, the addition of a non-polar compound such as toluene yields very moderate changes in the dielectric constants of the mixture even at high concentrations of toluene (Figure 6d)

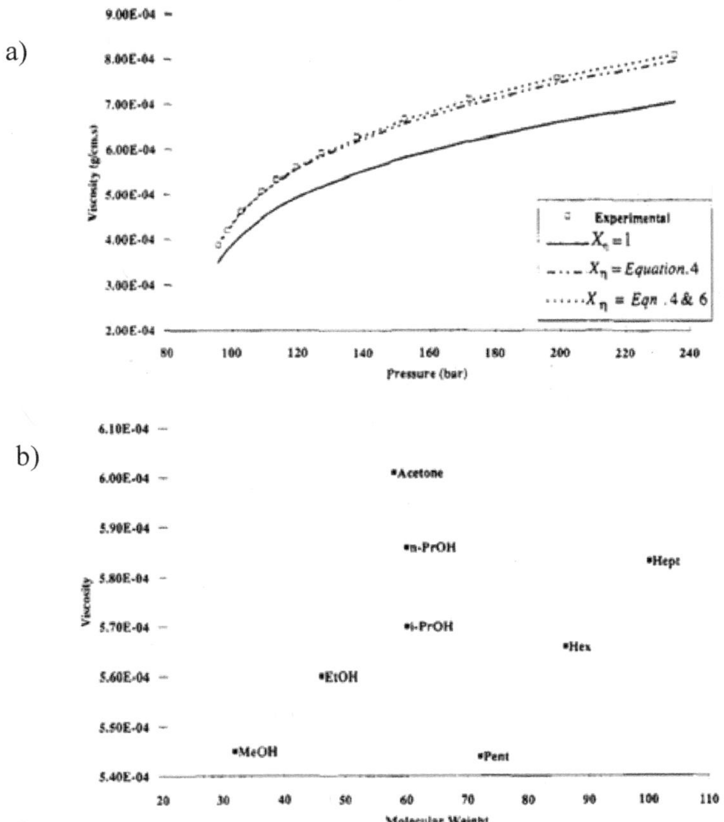

Figure 5. (a) Viscosity and density of 2% mol hexane in CO_2 at 45°C as a function of pressure, and (b) Viscosity of 2% mol co-solvent-CO_2 systems at 45°C and 120 bar as a function of co-solvent molecular weight. Reprinted with permission from K. D. Tilly et al. (Tilly et al. 1994) Copyright American Chemical Society.

(Wesch et al. 1996). Changing and controlling the dielectric properties of a solvent is of special interest in applications as varied as fabrication of polymeric fibers through electro-spinning (Luo et al. 2012; Park and Lee 2009), nanoparticles formation (Claverie et al. 2018) and assembly of structures (Rance and Khlobystov 2014), nanoparticles preparation by sol-gel reaction (Anwaar et al. 2019) and as fascinating and promising as proteins dynamics (Affleck et al. 1992).

Thermal conductivity of fluids is crucial in fields as varied as power generation, chemical production, automobiles, computing processes and refrigeration. Concerning solvents, regulation of thermal conductivity undoubtedly has an impact on engineering work involving heat transfer applications and design in chemical reactors, heat transfer equipment, and

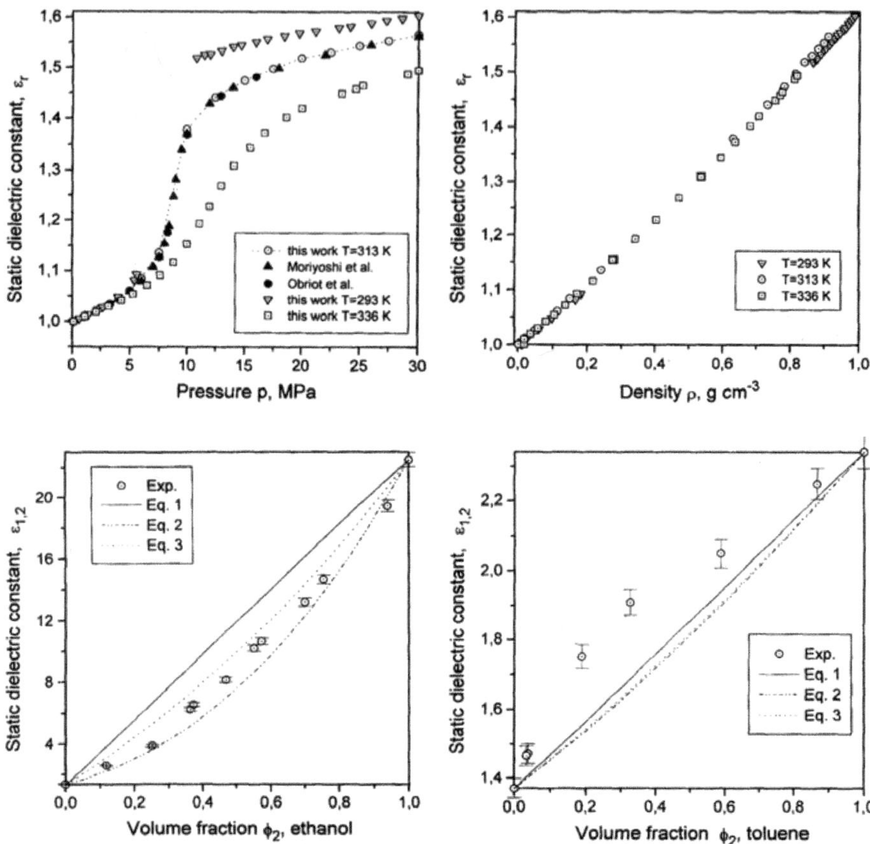

Figure 6. (a) Pressure dependence of static dielectric constants at different temperatures for pure CO_2, (b) Static dielectric constants of pure CO_2 as function of density at different temperatures, (c) Static dielectric constants of CO_2-etanol system at 10 MPa and 313 K as function of ethanol volume fraction, (d) Static dielectric constants of CO_2-toluene system at 10 MPa and 313 K as function of toluene volume fraction. Reprinted with permission from Wesch et al. (Wesch et al. 1996).

so on. The work of E. P. Sakonidou et al. (Sakonidou et al. 1996, 1998) and R. Mostert et al. (Mostert et al. 1990) examines the thermal conductivity of methane, ethane and a mixture 50/50 of these two gases, which illustrates the possibilities of modulation of this property at different temperatures and pressures when a mixture is used. It is to be noted that at reduced temperature differences $(T-Tc)/Tc > 10^{-2}$, the behavior of the thermal conductivity of a 50/50 mixture is similar to that of fluids with one component, where thermal conductivity continues to increase as temperature approaches Tc (Mostert et al. 1990; Sakonidou et al. 1996). However, for $(T-Tc)/Tc > 10^{-2}$ the thermal conductivity behavior of the mixture is completely different: it does not increase indefinitely but goes to a finite value (Sakonidou et al. 1998). This behavior has also been observed by D. G. Friend and H. M. Roder (Friend and Roder 1985, 1987). It is worthwhile to say that thermal conductivity data for mixtures SCF-co-solvent is scarce and, for most of the systems currently used, data is nonexistent.

From these works, it is clear that a modulation of solvent properties can be obtained by using SCFs and SCF-co-solvent mixtures. However, where concentration of the co-solvent goes beyond the miscibility limit, a second phase starts to appear, and what is called a gas-expanded liquid (GXL) may appear. In Solvent Engineering, gas-expanded liquids represent a great opportunity to modify the properties of a solvent medium by controlling temperature and pressure. These fluids will be presented and discussed in next paragraphs.

2.3 Gas-Expanded Liquids

A gas-expanded liquid is a fluid composed of a compressed gas and an organic solvent (Jessop and Subramaniam 2007). The most commonly used gas to generate a GXL is compressed CO_2 and the resulting fluid is often named CXL (CO_2-expanded liquid). The principle that allows Solvent Engineering by using GXLs is that by varying the operating pressure, the concentration of gas can be controlled, which allows obtaining a fluid media with physico-chemical properties tailored between those of the neat organic solvent and those of the pure gas (Jessop and Subramaniam 2007). This phenomenon can be reversed by depressurization which allows us to obtain the neat organic solvent of the beginning. A gas-expanded liquid presents an expansion when the gas solubilizes into the solvent; however, not all the solvents expand at the same extent when in the presence of a gas, namely CO_2. As the extent of the expansion of a CXL will be at the origin of the extent of the modulation of the physico-chemical properties of the expanded fluid, Jessop and Subramaniam (Heldebrant et al. 2006) have classified CXLs in three categories: Category I concerns solvents that solubilize CO_2 so poorly that expansion and properties change is not significant, an example of this category is water. Category II concerns organic solvents that dissolve great amounts of CO_2 and present a sometimes enormous expansion; most of the organic solvents fall in this category. Category III groups solvents that solubilize CO_2 to some extent and present great changes in some physico-chemical properties such as viscosity, but other properties such as volume do not change to a great extent. Some liquid polymers and ionic liquids fall in this category (Heldebrant et al. 2006).

Figure 7 schematizes the extension of the modulation of π^* Kamlet-Taft parameter of several solvents families as a function of pressure. These parameters reflect the polarity/dipolarizability and the ability to donate or accept hydrogen bonds of a solvent respectively. As can be observed, this

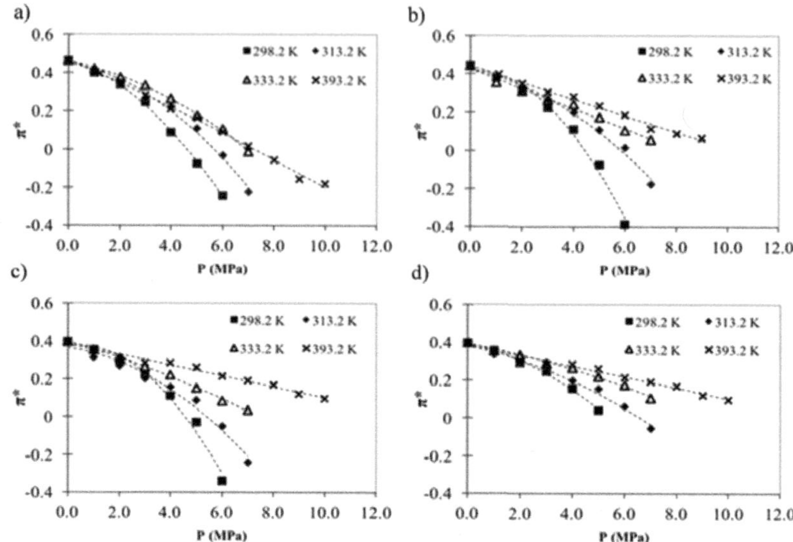

Figure 7. Extension of the modulation of π^* Kamlet-Taft parameter as a function of pressure at different temperatures for (a) methyl, (b) ethyl, (c) propyl, and (d) isoamyl acetates expanded by CO_2. Reprinted with permission from Granero-Fernandez E. et al. (Emanuel Granero-Fernandez et al. 2018). Copyright (2018) American Chemical Society.

modulation can confer intermediate properties to the mixture in a continuous and reversible fashion.

As for changes in properties such as density, concentration of CO_2 in the liquid phase is modified by pressure, changing the volume of the liquid phase, which naturally changes the density. In Figure 8, liquid-vapor equilibrium (LVE) at different temperatures of two systems formed by CO_2 and ethanol and CO_2 and acetone respectively is depicted: here, two phases can be observed (the CXL).

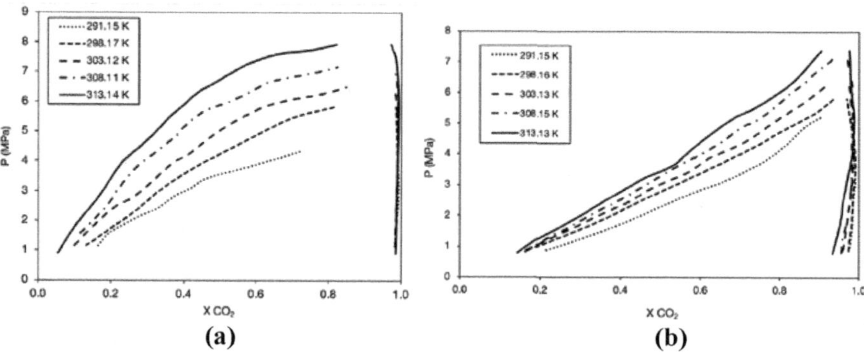

Figure 8. Liquid-vapor equilibrium for (a) CO_2-ethanol and CO_2-acetone systems at different temperatures. Data taken from reference (Day et al. 1999).

Given the tremendous possibilities that CXLs represent in Solvent Engineering, a good number of scientific reports have been published concerning the use of CXLs for extractions, chemical reactions, materials fabrication, and so on, and several excellent reviews have been published regarding these and other applications of CXLs. Interested readers are encouraged to consult them (Hintermair et al. 2010; Jessop and Subramaniam 2007; Musie et al. 2001; Subramaniam 2010b, 2010a, 2012; Subramaniam et al. 2014; Wei et al. 2002).

It must be noted that the use of CXLs needs a detailed knowledge of phase behavior and of the physico-chemical properties of the expanded phase. Some publications have addressed the study of phases behavior (Sánchez-Camargo et al. 2018) of macroscopic transport properties such as density and viscosity and of microscopic characteristic of the expanded phase around a solute molecule named the cybotactic region.

The techniques used for the characterization and study of these fluids are

1) Experimental techniques, by using visual observations through a high-pressure view cell to detect bubble and dew points and volume expansion, FTIR spectroscopy, solvato-chromic measurements and rheometry; and

2) Modeling, addressed through methods such as equations of state (EoS), molecular modeling, PC-SAFT, UNIFAC and so on (Abbott et al. 2009; Granero-Fernandez et al. 2018; Granero Fernandez et al. 2017; Hoang et al. 2017; Palafox-Hernandez et al. 2019).

It is to be noted that in the context of Solvent Engineering, CXLs can be seen as an example of a solvent easy to modify by a process variable such as pressure, and relatively easy to implement in industry. However, it is necessary to say that CO_2 decreases the polarity of the initial solvent, as CO_2 molecule is not polar. This is a disadvantage of CXLs when solubilization of a polar molecule is needed. Of course, usage of a less polar solvent can present advantages when an anti-solvent effect is needed, as required for some separations and extractions, etc. However, solubilization of highly polar solutes is not easy when using a CO_2-expanded liquid, even though elegant examples have been described in literature of applications of CO_2 to change the solvation properties of a solvent. J. Palafox-Hernandez et al. (Palafox-Hernandez et al. 2019) have studied two carbon dioxide expanded solvents, namely 1-octene and nonanal, which are solvents relevant for the hydro formulation of 1-octene at different pressures and temperatures through molecular dynamics simulations. In the first stage, Gibbs Ensemble Monte Carlo simulations were conducted to obtain the single component and the binary phase diagrams; then molecular dynamics simulations of the liquid phases were performed to obtain information about the structural and transport properties by using the LAMMPS package. Diffusion coefficients were calculated by using the Einstein relation from the slope of the mean-squared displacement of the molecules; while shear viscosity was calculated by using the Green-Kubo relation. The results thus obtained were compared with experimental data with good accuracy. Deviations from the Stokes-Einstein relation were observed to appear at high concentrations of CO_2 or at highly expanded volumes. The radial distribution functions were used to study the liquid structure in the expanded phases; nonanal exhibited dimeric structures at high concentrations of CO_2. The introduction of CO_2 leads to a decrease in the association behavior of 1-octene and of nonanal, which increases mass transport. Indeed, for both solvents, it was observed that the inverse of the shear viscosity is linearly dependent on the volume expansion of the solvent.

From the modulation capabilities of dissolved gases in solvents, the possibility to modulate the ability to create hydrogen bonds is one of their most interesting features, as hydrogen bonds creation is among the principal intermolecular interactions that drive a large range of phenomena from biological events to surface and supra-molecular chemistry. For example, complex levels of structuring in biologically active macromolecules as well as gelation phenomena and self-assembly of synthetic polymers into macrostructures are governed by hydrogen bonding (Yerushalmi et al. 2006). The interest to control hydrogen bonding is then clear. However even though this possibility is very interesting, these applications have not yet been developed by using gas-expanded liquids. In our opinion, the reversibility of hydrogen bonding modulation in GXLs can open fascinating opportunities in many diverse fields.

2.4 Switchable Solvents

Switchable solvents are defined as solvents that reversibly change physical properties abruptly as a consequence of a reversible reaction in response to an external stimulus (Pollet et al. 2011). The most used stimuli are temperature changes or addition/removal of a gas. In general, the reaction is reversible, which allows the changed solvent to be brought back to its initial state. In the context of Solvent Engineering, the possibility to manipulate the physico-chemical properties of a solvent in a reversible fashion opens the door to fascinating applications in chemistry, materials development, chemical reactions, and chemical processes development, among others.

Different switchable solvents have been described in literature, most of them can be classified into Switchable Polarity Solvents (SPS) and Switchable Hydrophilicity Solvents (SHS). SPS are solvents capable to change their polarity from a non-polar to a polar fluid, an ionic liquid, in the presence of CO_2. They were first described by Jessop in 2005 (Jessop et al. 2005). Secondary amines alone or mixtures alcohol/primary amines are some examples of SPS. The polarity of the final solvent can be tailored by a judicious choice of the system: alcohols and amines with short alkyl chains yield highly polar solvents, while long alkyl chains yield less polar solvents. As can be easily imagined, these features condition the miscibility of the final solvent with other solvents. Viscosity and melting point are also influenced by the chain length: alcohols with short chains give solid products, while long chains yield products with low melting points. SHS are solvents that, in the presence of CO_2, can switch from hydrophobic to water-miscible solvents. Examples of SHS are amidines, tertiary amines and bulky secondary amines. Switchable solvents have been used as reaction media and for extractions. Some examples are given in the following paragraphs.

Jessop et al. (Jessop et al. 2005) prepared a non-ionic mixture of DBU (1,8-diazabicyclo-[5.4.0]-undec-7-ene) and 1-hexanol and exposed it to one atmosphere of CO_2 at room temperature, which converted it into an ionic liquid. Degassing of CO_2 through bubbling of N_2 or argon allowed reversing

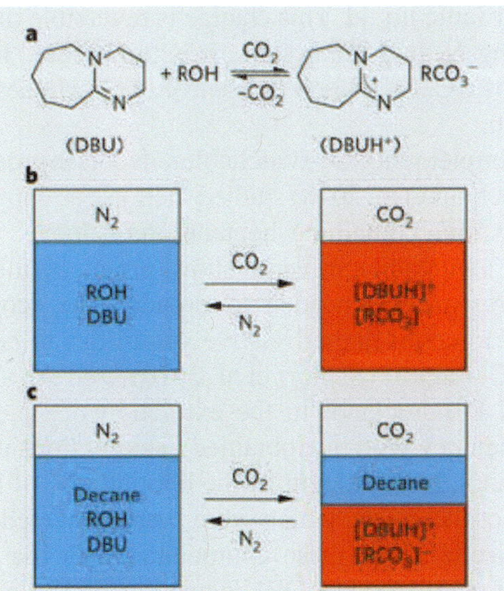

Figure 9. (a) Protonation of DBU in the presence of an alcohol and CO_2 is reversed when CO_2 is degassed. (b) CO_2 allows a non-polar liquid (hexanol and DBU) to change into a polar ionic liquid, degassing with N_2 allows this effect to reverse. (c) Changes in polarity are illustrated by tests of solubility of decane, which can be fully separated after addition of CO_2. Given the increase in polarity of the solution, full miscibility is recovered after addition of N_2. Taken from reference (Jessop et al. 2005) with permission from Springer Nature.

this process (Figure 9). Changes in polarity of the mixture (non-polar to polar and inverse effect) have been tested by observing its miscibility with decane.

Another elegant example of a switchable solvent is 2-butyl-1,1,3,3-tetramethylguanidine (TMBG) (Phan et al. 2008), which, in the same way as DBU, in the presence of CO_2 and an alcohol, forms an ionic medium. This process is reversed by bubbling an inert gas to evacuate the CO_2. Separation of products was easily triggered and achieved by using CO_2 to separate a mixture containing methanol, n-octane and the reaction medium. The products were solubilized in the octane phase and separated by decantation after adding of CO_2. Several cycles were performed using this system by including a drying step to eliminate the water formed during the reaction (Pollet et al. 2011).

Among the excellent examples of the possible utilizations of other SPS, there is the work developed by P. G. Jessop and E. Buncel (Boyd et al. 2013) where switchable non-ionic to ionic solvents were used to control the lifetime of merocyanine. Merocyanine is the open form of spiropyran. These two stable molecular states are inter-convertible through absorption of UV light, which is why they are called photo switches. Merocyanine lifetime has been observed to be enhanced in polar solvents. In this case, the switchable solvent tested is a mixture of DBU and alcohol, which, in the presence of CO_2, converts to a polar ionic liquid. This change is reversible through the removal of CO_2 by bubbling N_2 into the system. It is to be noted here that this work presents the use of CO_2 as a knob to control two switches: a solvent switch and a photo switch.

In an effort to implement CO_2-switchable solvents in processes engineering, B. Schuur et al. (Schuur et al. 2018) studied their use as entrainers in extractive distillation of mixtures containing heptane and toluene. The conclusion of this work is that switchable solvents show promises for separation steps. However, more research is needed to broaden the scope of compounds available in order to cover higher relative volatilities to be competitive with current solvents. C. Samorì (Samorì et al. 2010) used the switchable solvents DBU/ethanol and DBU/octanol in the extraction of lipids from algae and compared this efficiency with that obtained in extractions using n-hexane and chloroform/methanol. In this example, the lipophilicity of the non-ionic form was used to extract the lipids, which were then recovered by injecting CO_2, given the low miscibility of these compounds with the ionic form of the solvent (Samorì et al. 2010). The DBU/octanol switchable solvent was found to be more efficient for extraction than conventional solvents on dried and liquid samples, which allows envisaging the extraction without the drying step, which is a high energy consuming stage in chemical processes.

Extraction of soybean oil has been performed by using

1) a SPS (DBU/alcohol mixture) in order to use the low-polarity form to extract and the polarity-form to separate the oil,

2) a SHS, where firstly the SHS in its hydrophobic form was used to extract the soybean oil which had been exposed to water, and then the hydrophilic form was used to separate the solvent/water mixture from oil and

3) dioxane, to extract oil from soybean flakes, and then the mixture was washed with water and exposed to CO_2 in order to induce a phases separation.

Among these three strategies, the second one was the most reliable by using an amidine, as only small quantities of residual amidine were found in the oil obtained (Phan et al. 2009).

Examples of uses of CO_2 as a trigger in switchable solvents, surfactants and other materials published up to year 2012, are gathered in the review by P. G. Jessop et al. (Jessop et al. 2012). More recent examples of switchable solvents concern switchable ionic liquids based on mixtures of DBU, glycerol and acid gases such as CO_2 or SO_2 that have been proposed by Anugwom et al. (Anugwom et al. 2011). It is interesting to observe that the inclusion of glycerol increases the decomposition temperature of these solvents, as has been proven by DSC studies, which makes these solvents specially interesting for industrial applications. In this study, the switches used were CO_2 or SO_2, and an increase of viscosity was observed after the switching.

In the domain of extraction of aromatic hydrocarbons and alkanes, molecules synthesized by introducing OH^- ions in molecules derived from the amidine compound DBU have been used for the separation of toluene from *n*-heptane used as a model hydrocarbon mixture (Zhang et al. 2020). It has been concluded that the hydrocarbon chain length of the molecule derived from DBU played a key role in selectivity towards toluene, and the original switchable solvent was recovered when heated at 60°C for 1 h. The NRTL model was used to correlate and predict the experimental results obtained. Another interesting work concerning the use of DBU, methanol and CO_2 for the synthesis of methyl propionate and methyl methacrylate without catalysts was published by S. G. Khokarale and J.P. Mikkola (Khokarale and Mikkola 2019), where methanol was finally efficiently separated from the reaction mixture along with the products obtained. Water-immiscible fatty acids were switched to water-soluble by complexation with amines and CO_2-bubbling.

The use of CO_2 to trigger a change in solvents and then a separation in the phase can be judiciously used in separations, phases transfer catalytic reactions, and so on. In this sense, this strategy is included in this book, and in this section devoted to switchable solvents, even though a phase's separation by using CO_2 as a trigger has been presented as a strategy for CO_2 capture too. Here, a one-phase fluid is held in contact with CO_2 thus becoming two immiscible liquid phases, one CO_2-rich and the other CO_2-lean, where the CO_2-lean phase is sent back to the process. The energy consumed by this operation is drastically reduced compared with conventional post combustion carbon capture processes. Some systems include an amine dissolved in an alcohol. Here, the CO_2-lean phase is almost the alcohol with traces amounts of amine, while the CO_2-rich phase is mainly amine-bonded CO_2 and unreacted amine (Machida et al. 2018). It is interesting to observe that energy consumption of the solubility of CO_2 in a phase separation solvent has been compared to the solubility in a miscible and in an immiscible solvent with the conclusion that the process including a phase separation solvent is more favorable than the other two if the former includes a heat pump to decrease the use of energy (Machida et al. 2018).

F. Liu et al. (Liu et al. 2021) have developed CO_2-switchable deep eutectic solvents (DES) formed by imidazole as the hydrogen bonding acceptor and a polyol such as ethylene glycol, glycerol, 1,4-butanediol and polyethylene glycol as the hydrogen bonding acceptor. Reversible emulsions and phases separation were obtained by bubbling CO_2 or N_2 into a solution of olive oil and the imidazole-based DES: the emulsion was destroyed by CO_2 injection, while the emulsion was recovered after N_2 contact. Evidence of the formation of carbamate after bubbling of CO_2 was found by ^{13}C NMR. Interestingly, conductivity of the system increased through the formation of anions and cations between CO_2 and imidazole.

Recently, R. R. Benedix et al. (Benedix et al. 2023) have prepared a mixture of tetradecyltrimethylammonium bromide ($C_{14}TAB$) and N,N,N,N-tetramethyl-1,4-butanediamine (TMBDA), which is a CO_2-switchable molecule. By switching the TMBDA with CO_2, a control of the surface and tensio-active properties of the mixture could be obtained modifying its foamability and foam stability, which can be judiciously used in applications in mineral flotation. In this study, the authors affirm that switching TMBDA to its diprotonated form by CO_2 changes the surface composition, which induces the changes observed in surface tension. Diprotonated TMBDA stays in the bulk and is not surface-active while TMBDA is somewhat surface-active and destabilizes the surfactant action. Foams stabilized by $C_{14}TAB$ alone exhibited more stable behaviors than foams formed by CO_2 in the system comprising $C_{14}TAB$ and TMBDA.

A review about CO_2 phase separation absorbents has been published by Q. Zuang et al. (Zhuang et al. 2016). This strategy is presented here as changes in solvents induced by CO_2 represent change in solvency and transport properties of the involved solvents too, which can be used for applications other than CO_2 capture.

2.5 Mixtures of Solvents

Optimizing a chemical or physical process performed within a solvent by adjustment of the properties of the solvent medium has been performed for long time by mixing two or more solvents. Comprehension and eventually prediction and control of the properties of a mixture of solvents has attracted great attention from researchers in physico-chemistry and thermodynamics for over a century. A good number of theories and models have arisen from this interest, aiming at better understanding solvency phenomena when mixtures of solvents are used to better control the properties of this mixture.

Intuitively, a mixture of, for instance, two solvents, would present intermediary properties between those of the pure solvents that form the mixture. However, some interesting surprises can be found in mixtures of solvents where the properties of the resulting mixture do not exhibit the expected properties. One example of intriguing behavior can be found when the mixture of two miscible good solvents for a polymer becomes a poor solvent for the same polymer. This behavior was explained by D. Mukherji et al. for a mixture of methanol or ethanol and water which separately are good solvents for poly(N-isopropylacrylamide) (PNIPAm), but precipitate in mixtures of these solvents (Mukherji et al. 2014). These authors have shown through numerical simulations that the preferential adsorption of one of the solvents onto the chain backbone controls collapsing and re-swelling of the polymer. From this study, it has been shown that the origin of these complex co-solvent-driven conformational transitions is generic, meaning that a broad range of polymeric systems can show similar behavior. This study shows an interesting way to control transitions from globular to coiled structures of polymers through control of the concentration of solvents in the mixture containing the polymer.

Control of intermolecular interactions solvent-solvent, and solvent-solute has attracted great attention recently with the increasing interest in self-assembly in chemistry and in soft matter. Among the most prominent non-covalent interactions, hydrogen bonding (H-bonding) is one of the strongest and most directional of intermolecular interactions. H-bonding interactions are very sensitive to solvent environment,

In a simplistic fashion, polar interactions are favored by non-polar solvents, while attractive interactions between non-polar groups are obtained in polar solvents. On the other hand, preferential solvation is a phenomenon where solvent proportion in the vicinity of a solute inside a solvent mixture differentiates from the statistic proportion in the bulk, which is called the solvation shell or the cybotactic region (Figure 10) (Morisue and Ueno 2018). This preferential solvation phenomenon can be used to characterize solvent environments or to judiciously modify the behavior of certain solutes. In all

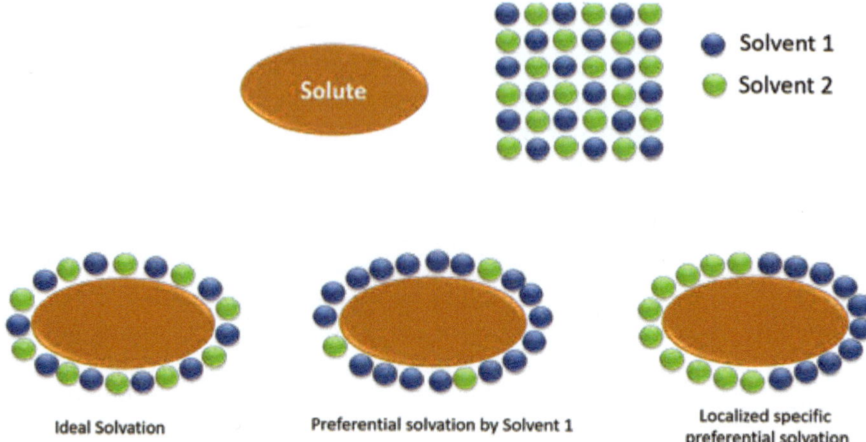

Figure 10. Schema of the arrangement of two solvents in the cybotactic region of a dipolar solute (blue) in a 1:1 binary mixture of solvents 1 (blue) and 2 (green). Case of (i) ideal solvation. (ii) preferential solvation by solvent 1, and (iii) localized specific preferential solvation. Adapted from Malik et al. 2020.

cases, preferential solvation needs to be taken into account when, for instance, non-covalent interactions are present in a solvent mixture.

Preferential solvation by specific H-bond interactions of the probe tri-*n*-butylphosphine in binary solvent mixtures has been evidenced by Cabot et al. through measurements of ^{31}P NMR chemical shift as a function of solvent composition. Tri-*n*-butylphosphine is one of the best hydrogen-bond acceptors in a competitive polar solvents environment, allowing this strategy to probe solvent environments. The results obtained have been related to spectroscopic solvent polarity scales, showing that this method is useful to measure non covalent effects of mixtures of solvents in solutes (Cabot and Hunter 2010). In another study, J. L. Cook et al. (Cook et al. 2007) performed experimental measurements of association constants for the formation of complexes of perfluoro-*tert*-butyl alcohol, which is one of the most polar hydrogen-bond donors and tri-*n*-butylphosphine. In this study, the authors used ^{31}P and ^{19}F NMR signals of these compounds respectively to obtain information about complex formation. The ^{31}P NMR signal of tri-*n*-butylphosphine is particularly sensitive to hydrogen-bonding interactions. This strategy was later used by the same authors (Cook et al. 2008) to study the preferential solvation and hydrogen bonding in mixed solvents, by using tri-*n*-butylphosphine oxide and perfluoro-*tert*-butanol, which complex is presented in Schema 1. This complex was used as a probe of solvent effects on hydrogen bonding interactions in mixtures of chloroform – tetrahydrofuran. This mixture does not follow the regular solution theory and shows strong deviations from ideality; for instance, addition of chloroform (a less polar

Schema 1. Tri-*n*-butylphosphine oxide and perfluoro-*tert*-butanol complex. Adapted from Cook et al. 2008.

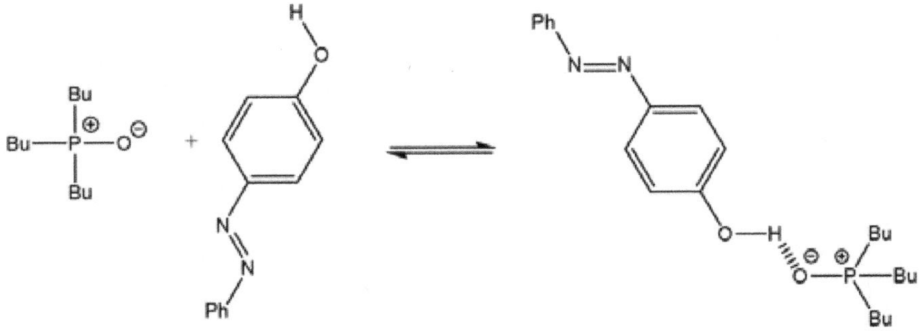

Schema 2. Formation of a 1:1 complex between the tri-*n*-butylphosphine oxide and the 4-phenyl azophenol.

solvent) to tetrahydrofuran (a polar solvent) produces a significantly more polar mixture. This interesting feature allows the use of this and other strongly non-ideal mixtures in Solvent Engineering.

Other mixtures that present complex solvation environments have been studied by the same group (Amenta et al. 2013), and concerned the role of solvent self-association on the solvation properties of polar liquids. For this study, mixtures of alcohols, carboxylic acid and secondary amides were studied by using the complex formed by a H-bond donor (4-phenyl azophenol) and a H-bond acceptor (tri-*n*-butylphosphine oxide) (Schema 2).

This same group has developed a tool to quantitatively predict how a solvent or a mixture of solvents will affect non-covalent interactions in solutes, referring to it as 'functional group interaction profiles' (FGIP) (Driver et al. 2020). The FGIP analysis allowed the authors to propose a quantitative tool for the understanding and prediction of non-covalent interactions between solutes with their solvent environment by representing the molecular surface as a discrete set of surface site interaction points (SSIP). For example, by taking into consideration the case of a mixture of chloroform and tetrahydrofuran, it is proposed that preferential solvation leads to good solvation of hydrogen bond donor solutes by tetrahydrofuran, while good solvation of hydrogen bond acceptor solutes will be achieved by chloroform. As result, the two pure solvents will favor the interactions between polar solutes while mixtures will disadvantage these interactions. In the case of an ethanol–water mixture as well, ethanol is a solvent that owns both polar and non-polar surface SSIPS. Consequently, in ethanol, non-polar solutes can choose between interactions with the polar SSIPS associated with the OH group or the non-polar SSIPS associated with the ethyl group. In water, non-polar solutes can only interact with the polar SSIPS leading to poor solvation and favorable solute-solute non-polar interactions, i.e., the hydrophobic effect. The hydrophobic effect is less pronounced as the amount of ethanol in the mixture increases and disappears at 100% of ethanol. The authors have calculated FGIPs for about 300 different solvents and solvent mixtures that allow an interesting analysis adapted to perform Solvent Engineering to optimize non-covalent interactions during, for instance, the fabrication of materials of interest with controlled conformations.

Preferential solvation of the solvato-chromic probe p-nitroaniline (PNA) in mixtures chloroform–hydrogen bond acceptor solvents has been studied by P. Kumar Malik et al. (Malik et al. 2020) by using spectroscopic transition energy (E_T). The authors used the Bosch and Rosés model to address the role of specific interactions between solute and solvent: this model has been used to describe the solvation of probe molecules in binary mixtures of solvents. The model proposes to calculate preferential solvation parameters from the bulk mole fractions of the solvents in the mixture and from the mole fractions of the solvent in the cybotactic region of the solute, which could be obtained from the Bagchi and Chatterjee model and experimental determinations of E_T.

G. N. Pallawela and P. E. Smith have used the Kirkwood–Buff theory that relates thermodynamics of solutions to the distribution of molecules inside a liquid mixture by using the Kirkwood–Buff integrals (Pallewela and Smith 2015). This approach is an extension of the Ben-Naim approach, which was the first to provide a rigorous framework to study preferential solvation in mixtures through the use of experimental Kirkwood-Buff integrals. In the work presented by G. N. Pallawela and P. E. Smith, the Ben-Naim treatment is completed by including the desired properties of an ideal mixture. According to them, experimental data such as density, activity and compressibility can be used to obtain these integrals and so to give a quantitative idea of the affinity between the compounds of a mixture. So, thermodynamic properties can then be related to preferential solvation and to the structure of the solution. In particular, the authors have studied mixtures of methanol with water, methyl acetate and 1,2-dichloroethane by using the GROMACS simulation package to perform classical molecular dynamics simulations of these mixtures and to obtain information of the spatial dependence of preferential solvation. According to the authors, the approach and the thermodynamic treatment presented in this study is extensible to ternary mixtures.

The study and even engineering of the preferential solvation of drugs in mixtures of solvents is important for their production, separation and formulation as, very often, drugs are only partially soluble in water. Y. Marcus (Marcus 2017) has presented a study of the preferential solvation of some drugs in binary mixtures of solvents by using the inverse Kirkwood-Buff integrals and the quasi-lattice quasi-chemical methods and data from literature. For instance, in the case of caffeine and niflumic acid in mixtures of ethyl acetate and ethanol, it can be concluded that both drugs have a preference for ethanol at low concentrations, while at higher concentrations, both of them prefer ethyl acetate. In the case of mixtures of ethanol/water, niflumic acid prefers ethanol in all solvent concentrations; caffeine prefers ethanol at low ethanol concentrations, while at high ethanol concentrations, caffeine prefers water. The study of preferential solvation has been the object of a multitude of studies. The understanding, modeling and eventually the control of this phenomenon falls within the scope of Solvent Engineering as it allows for improving and optimizing the solvency and transport phenomena occurring in the solvent medium.

Another class of important, directional intermolecular interactions are halogen bonds (X-bonds), denoted as R-X⋯Y ('R-X' and 'Y' denote the halogen bond donor and acceptor respectively), which involve the interaction between a positively charged cap (called a σ-hole) along the axes of the R-X covalent bond and a nucleophilic part of a molecule or atom (Wu et al. 2018) containing lone pairs (Lewis bases). Hydrogen and halogen bonds formation can become a competition and can be controlled by a solvent medium, thus influencing phenomena in self-assembly. This enormous effect of solvents was experimentally evidenced by C. Robertson et al. (Robertson et al. 2017) through experiments of co-crystallization, showing the critical effect of the solvent medium in self-assembly, and that careful choice of solvent results in either hydrogen-bonded or in halogen bonded co-crystals. Another interesting example where halogen and hydrogen bonding was evidenced and used to develop a 2D self-assembly system through a judicious choice of the solvent was presented by Y. Wu et al. (Wu et al. 2017). In this study, the authors observed the 2D self-assembly of a bi-functional molecule: 5-bromo-2-hexadecyloxy-benzoic acid (5-BHBA) in two different solvents: 1-octanoic acid and n-hexadecane at a solid-liquid interface. Here, cooperative and competitive halogen and hydrogen bonds were related to the polarity of the solvent. By adjusting the solution concentration of 1-octanoic acid and n-hexadecane, different kinds of self-assembled patterns were obtained, allowing the control of the structures obtained by concentration of the solvent mixtures.

An elegant example of the solvent effect on mechanical motion in supra-molecular systems such as rotaxanes has been presented by P. Ghosh et al. (Ghosh et al. 2005). In this study, the speed of rotation of the prepared rotaxanes has been reversibly modulated from 'slow' to 'fast' through the modulation of the electrostatic interactions with the solvent, which has been induced by altering the proportion of polar solvent in a binary solvent mixture, for instance, concentration of DMSO in dichloromethane.

Deep eutectic solvents (DES)-water mixtures can be used to control the associating behavior of anionic or cationic surfactants such as sodium dioctyl sulfosuccinate, and sodium dodecyl sulfate, as has been observed by K. G. Singh et al. (Singh et al. 2018), M. Pal et al. (Pal et al. 2014; Pal et al. 2015), A. Sanchez-Fernandez et al. (Sanchez-Fernandez et al. 2016), A. Yadav and S. Pandey (Yadav and Pandey 2014) and R. B. Leron et al. (Leron et al. 2012), among others. It is interesting to note that DES-water mixtures present heterogeneities and so nano segregation of solutes is possible, which induces specific interactions, and thus, aggregation behaviors are different when neat DES or their mixtures are used. For instance, sodium dioctyl sulfosuccinate forms aggregates similar to micelles in neat DES such as those formed by choline chloride + urea (reline) and choline chloride + ethylene glycol (ethaline) while nano segregation of the surfactant is observed in mixtures formed by DES and water.

A. Nag et al. (Nag et al. 2022) have proposed the use of mixtures of triphenylphosphine dichloride or of oxalylchloride with hydrogen peroxide for the solubilization of elemental noble metals such as gold, platinum, palladium, and copper, yielding metal chloride salts, allowing us to envisage a recycling procedure from electronic wastes without the use of strong acids and/or more toxic reagents. Biphasic mixtures were used during this work to precipitate the solubilized metals. Concerning the recovery of metals from printed circuit boards such as copper, tin, zinc, iron, and nickel, S. Mishra et al. (Mishra et al. 2024) have proposed a method using choline chloride (ChCl) based deep eutectic solvents, obtained by mixing ChCl with formic acid or urea, after metal leaching with H_2O_2: metal recovery was more than 90% at optimized conditions. In another study, DES formed by monochloroacetic acid (MCA) and tetrabutylammonium chloride (TBAC) was used to recover the metals contained in lithium-ion batteries such as Li, Ni, Mn, and Co. To preserve the DES from damage caused by the metals, oxalic acid was added. This compound was used to obtain leaching of the metals after their solubilization. Recycling of the solvent was possible after the leaching process and evaporation (Zhang et al. 2024).

A compilation of experimental work and data about properties and preferential solvation in mixtures of solvents has been published by Y. Marcus (Marcus 2002). Interested readers are advised to consult it for a deeper insight of the thermodynamics and physico-chemical properties of mixtures of solvents.

2.6 Supramolecular Solvents (SUPRASs)

Supramolecular solvent (SUPRAS) is a term adopted recently for water-immiscible liquids made up of supramolecular assemblies dispersed in a continuous phase. SUPRASs are obtained from amphiphile solutions by a two-step self-assembly process occurring at two scales, molecular and nano (Ballesteros-Gómez et al. 2009). This process first gives three-dimensional aggregates that coacervate, and, in the second stage, produce water immiscible liquids made up of large supramolecular aggregates dispersed in a continuous phase, generally water. It is to be noted that the formation of a supramolecular solvent involves non-covalent interactions through which molecules are held together in the solvent, and self-assembly processes through which they are formed.

Briefly, a SUPRAS is formed after a first self-assembly process of amphiphilic molecules. These amphiphile molecules associate when the critical aggregation concentration (cac) is reached in order to minimize unfavorable solvophobic interactions. There exist at this stage, however, head group-head group repulsions that stop aggregation; after this step, the amphiphilic aggregates self-assemble through coacervation and produce an amphiphile-rich liquid phase, which reduces the repulsions that stopped the aggregation in the first stage (Ballesteros-Gómez et al. 2009).

The solvent properties of SUPRASs can be tuned by proper selection of the amphiphiles and of the environment for their self-assembly (Caballo et al. 2017). Indeed, to accomplish the formation of a supramolecular solvent and to control its solvation properties, different variables can be modified, such as the size and nature of the hydrophilic and the hydrophobic parts of the amphiphile, addition of amphiphilic counterions or electrolytes, modification of temperature, changes in pH and addition of a poor solvent. Temperature has been the most common inductor of coacervation in colloidal solutions of non-ionic, zwitterionic, and mixtures of nonionic and nonionic/ionic surfactants. Up to now, supramolecular solvents applications can be found in the field of extractions for analytical chemistry and for the recovery of traces of metals (Ballesteros-Gómez et al. 2010; Bezerra et al. 2005; Melchert and FábioRocha 2016; Melnyk et al. 2015; Ojeda and Rojas 2009; Paleologos et al. 2005; Samaddar and Sen 2014; Silva et al. 2006; Stalikas 2002; Urík 2016).

The development of supramolecular solvents should grow in future years, as supramolecular chemistry is a mature discipline and a good amount of information and knowledge about coacervates formation, self-assembly and molecular interactions has evolved in recent years. However, this information is largely spread among the colloid, polymer, physical chemistry, analytical and pharmaceutical literatures. Greater efforts need to be addressed to the synthesis of these research works for the systematization of the use of these solvents and to develop new supramolecular solvents. In this regard, usage of Artificial Intelligence, machine learning and other computing techniques can help in the systematic analysis of scientific literature and in the prediction of properties and of new supramolecular solvents, as well as help in their utilization in current and new application fields.

2.7 Liquid Crystals

The use of liquid crystals (LCs) as solvents is not new. Their use has been periodically reviewed as reflected in the chapter by W. J. Leigh and M. S. Workentin (William et al. 1998) concerning their use as solvents in spectroscopy, chemical reactions and gas chromatography. The use of different phases of liquid crystals allows interesting separations (Dewar and Schroeder 1964). The synthesis and use of ionic liquid crystals (ILCs) was recently proposed (Suisse et al. 2005). ILCs are materials that merge the properties of ionic liquids and of liquid crystals. A thorough review of ILCs has been published by K. Goossens et al. (Goossens et al. 2016). From these review works, it can be said that the research performed to understand and synthesize ILs can be used for the design and application of LCs and of ILCs.

G. Celebre et al. (Celebre et al. 2005) worked on the study of chiral induction in liquid crystals, and more precisely the induction of a cholesteric phase through interactions between solvent and solute during the phase formation of methyl phenyl sulfoxide. During the formation of the structure in nematic solvents, cholesteric phases of opposite handedness were obtained. An extensive liquid crystal NMR (LXNMR) analysis was used to study the conformational arrangement of solutes in the liquid crystals and to obtain the conformational distribution and orientational order of the dopant. The authors have shown that changes of cholesteric handedness with a solvent are substantially driven by changes in the orientation of the dopant.

N. R. Scruggs et al. (Scruggs et al. 2006) proposed the use of a liquid crystal as switchable solvent to mediate the segregation between coil and liquid crystalline polymers. Here, a liquid crystal was used as solvent to induce the self-assembly of a side-group liquid crystalline polymer. The rheology of the solutions was correlated to the self-assembly of the polymer obtained when the solvent was changed from a nematic to an isotropic phase. In the nematic phase of the solvent, elastic gels were obtained from the strong interactions between dense-core micelles; this elastic behavior decreased when the solvent started to enter the micelles upon isotropization of the solvent, resulting in a transition from a viscoelastic to a viscous liquid behavior. During the nematic phase, selectivity of the solvent drives the self-assembly, while during the isotropic phase, the selectivity decreases and the micelles formed swell up to their complete solubilization.

P. C. Mushenheim et al. (Mushenheim et al. 2016) studied the mechanical strain produced on giant unilamellar vesicles confined by aqueous chromonic liquid crystal phases. By producing thermal quenching from isotropic to liquid crystal phases in the solvent, the vesicles went from spherical to two distinct populations with elongated spindle-like shapes. The liquid crystal used as solvent was a 15% (wt/wt) solution of disodium cromoglygate (DSCG) that, at this concentration, formed a nematic phase below 30°C and an isotropic phase above 42°C.

To understand the effect of the anisotropy of a solvent, for instance a liquid crystal, on electron transfer activation reactions, M. Lilichenko and D. V. Matyushov (Lilichenko and Matyushov 2003) performed Monte Carlo simulations of the distribution of the fluctuations of the solvent polarization at the microscopic scale in a range of densities that covered the nematic to the isotropic phases of the solvent. The model fluid used was a fluid formed by hard sphero-cylinders presenting a dipole moment. The reduced rate constants of a model reaction from a non-polar state (charge separation) to a polar state (charge recombination), were calculated and related to the quality of the solvent.

Ionic liquids with liquid crystal properties have been prepared by some research groups and reviewed by T. Ichikawa et al. (Ichikawa et al. 2019). These compounds are called ionic liquid crystals. One of the strategies to prepare these ionic liquids is to introduce a ionophobic long alkyl chain into the ions to generate nano segregated states that act as a driving force to self-organization into dimensionally ordered states. For instance, imidazolium and pyridinium salts with a long alkyl chain on their cations can form smectic phases formed by ionophilic and ionophobic two-dimensional layers. Protic ionic liquid crystals can be obtained by the neutralization of acids and bases, for instance, those prepared by hydrogen bonding between pyridine and carboxylic acid derivatives or by the neutralization of n-alkylamines and aromatic acids. Some amino acids-based ionic liquids have shown liquid crystal phases in the presence of amphiphilic molecules; the structures formed can be modified by using divalent or aromatic amino acids. The authors have suggested that the strategy of designing new controlled solvents for desired structures instead of designing new amphiphiles may be promising for the design of new structures in soft matter.

Smectic solvents have been proposed as reaction solvents for the synthesis of bicyclic cycloadducts by intramolecular uncatalyzed cycloaddition of 2E,7E,9(E)-decatrienoic and 2E,8E,10(E)-undecatrienoic acid esters as probe reaction by K. Fukunaga and T. Kunieda (Fukunaga and Kunieda 1999). The anisotropic rigidity of the solvent was found to be the key for the enhancement of diastero control and chemical efficiency. Degradation of the products was observed in isotropic phases while an acceleration of the cyclo addition reaction was clearly observed in the anisotropic phase. In conclusion, the smectic phase rigidity allowed control of the reaction with good diastero selectivity and chemical efficiency.

Indeed, fine tuning of the phase behavior and of the temperatures of transition as well as the rheological properties of these materials can be used for the design of optimized anisotropic solvents, which can be used as organized reaction media inducing an orientation and a confinement of reagents, which can serve to control regio-, stereo- and/or enantio selectivity in some reactions and self-assembled systems.

2.8 Active Fluids

Active fluids are a wide class of non-equilibrium systems consisting of suspensions of interacting self-propelled particles, cells or macromolecules able to convert the energy stored in the environment into mechanical work. These fluids, by continuously consuming energy, give rise to non-equilibrium thermodynamic effects that are not observed in conventional fluids. Different mechanisms can be present that induce a spontaneous flow of particles in a fluid, for instance, a chemically reacting boundary layer as in autophoretic colloids, a dissolution layer like in auto-osmophoretic drops, propagation of waves along cilia or flagella in synthetic swimmers, self-generated gradients of temperature in thermophoresis-propelled particles, or gradients of electric potential as in electrophoresis-propelled particles, among others. Possible applications of such fluids have been evoked in biomedical and engineering applications such as cargo transport for targeted drug delivery and in micro-machines (Thutupalli et al. 2018; Varma et al. 2018).

Active particles influence each other's movements through chemical and hydrodynamic interactions which give rise to self-organization and collective behavior. It is interesting that treatment of these particles as a continuum can give rise to bulk properties that can be treated as the characteristics of a solvent matrix, and that concern transport and solvency properties. From a Solvent Engineering point of view, self-organization and collective behavior of active particles can be useful for tuning active materials properties and to induce some pre-defined behavior. In this context, and from a transport properties point of view, treatment of active colloids or particles as a continuum has been proposed by some authors, and can be found in literature (Caviglia and Morro 2015; Pedley and Kessler 1990).

Recent studies, most of them theoretical, reveal fascinating behaviors of active suspensions, for instance enhanced mixing and self-diffusion, reduction of the effective viscosity, active clustering, artificial rheotaxis and light controlled diffusion and flows, among others (Ebbens 2016; Palacci et al. 2014; Saintillan 2018). Such a control of properties as desired would be of great use in Solvent Engineering.

Solvent effects on active particles are crucial. For instance, simulations performed with Janus and dimeric colloidal swimmers (Yang et al. 2014) have explored precise solvent behavior and show a fundamental different hydrodynamic behavior. Mesoscopic computer simulations, and in particular the use of a particle-based method known as Multiparticle Collision Dynamics (MPC), in which the solvent is explicitly considered, is a powerful tool when considering modeling of active particles. A. Bera et al. (Bera et al. 2022) have studied systems containing active particles in an explicit solvent. They have taken into consideration the hydrodynamics of the solvent by using a hybrid simulation method, using molecular dynamics and MPC models to observe the dynamics of the clustering of the particles to study phase separation of active matter systems. Density of the solvent influenced the movement of the particles either in the passive or in the active state. A. Varma et al. (Varma et al. 2018) have studied the self-propulsion of autophoretic particles induced by clustering. In this study, the authors modeled phoretic and hydrodynamic interactions to observe the clustering arrangement of particles during dynamic phoretic clustering, for which hydrodynamics was thoroughly studied.

In the same sense, active fluids may present curious rheological behavior, as has been detailed in an article presented by L. Giomi et al. (Giomi et al. 2010). They have studied the rheological behavior of thin films of polar and non-polar active particles, concluding that for weakly active systems, activity can decrease the linear bulk viscosity of tensile suspensions of swimmers and enhance the viscosity of contractile systems. They found that the form of the particles also plays a role. While activity lowers the viscosity of both tensile, rod-shaped particles and contractile, disk-shaped particle suspensions, it increases the linear viscosity of contractile, rod-shaped particle and tensile, disc-shaped particle suspensions. For strongly active systems, the rheological behavior is strongly non-linear. They have proposed some explanations for that, including the existence of a superfluid phase.

Newtonian or non-Newtonian behavior of the solvent influences the Brownian behavior of particles, and changes in viscosity and density. Some authors have suggested that adding active colloid particles to a Newtonian fluid may provoke a complex fluid behavior.

From these and other studies not discussed here, it is clear that solvent properties play a role in active matter behavior. Controlling solvent properties then becomes a knob to control active matter behavior too, and, going further, it may be said that an active suspension, or a suspension of active particles, can behave at the macro-scale as a solvent, whose active properties can be controlled and used for innovative processes. It is however striking that these studies take into consideration solvent interactions with active particles only as hydrodynamic friction and as a means to transfer momentum, the chemistry of the solvent has never been taken into consideration in active particles studies, and neither have mixtures of solvents have been considered, which could open great perspectives in the active particles field.

2.9 Porous Liquids

Porous liquids are materials that present permanent porosity combined with fluidity. They were first described by James et al. in 2007 (James 2016; O'Reilly et al. 2007), who proposed to classify them into three types: Type I porous liquids are micro-porous molecules that maintain an intrinsic porosity in the liquid state, they exhibit a single fluid phase. Type II porous liquids are molecule-size species that have intrinsic porosity dissolved in a solvent medium. Type III porous liquids are composed of particles of micro-porous solids such as Metal Organic Frameworks (MOFs) or zeolites in colloidal systems, dispersed throughout bulky solvents size-excluded from the pores.

Porous liquids offer multiple opportunities. From a processes point of view, their fluidity allows continuous cyclic processes that are not possible when solids alone are used, and their porosity can enable size or shape-selective dissolution of solutes. Porous liquids can be engineered to obtain a specific composition or property, for instance, gas uptake, selectivity for separations, and so on.

Type III porous liquids have been used for the separation of ethane/ethene mixtures (Lai et al. 2021) as their separation is challenging because of the similar properties and sizes of both compounds Selectivity, gas uptake and regeneration were investigated for five porous liquids formulated in sesame oil or in an ionic liquid. The solids were chosen among zeolites and MOFs. Interestingly, one of the conclusions of the authors is that the gas uptake behavior of MOFs can be transferred to porous liquids. For very recent reviews of applications of porous liquids and the strategies to produce them, interested readers can access those published by B. D. Egleston et al. and by D. Wang et al. (Wang et al. 2022). T. D. Bennett et al. (Bennett et al. 2021) and H. Mahdavi et al. (Hamidreza Mahdavi, Smith et al. 2022), published comprehensive reviews including some guides for the creation of new porous liquids and provided perspectives on applications for porous liquids. According to these authors, the main challenges concerning porous liquids remain costs and the judicious design and engineering of the system solid – solvent medium. In particular, in a porous liquid, a preponderant role is played by the solvent. Here, one can consider either performing Solvent Engineering by using the porous liquid as an engineered solvent, or performing Solvent Engineering to obtain a particular porous liquid with tailored properties. H. Mahdravi et al. (Hamidreza Mahdavi, Zhang et al. 2022) have reviewed the solvent-based factors that influence permanent porosity in porous liquids, with the focus on steric hindrance that prevents solvents from entering the cavity of the porous solid. For this, zeolitic imidazolat frameworks were combined with different solvents and the capacity for CO_2 sorption of the solid was measured as the parameter to be optimized.

In another study, C.W. Chang et al. (Chang et al. 2022) have proposed a method to accelerate the selection of solvents for obtaining porous liquids of type II by using COSMO-RS calculations and pKa values of the solvent together with predictions by machine learning models. The obtained results were confronted with experimental data. A library of more than 11,000 compounds as solvents was used in this approach to select the most appropriate to obtain type II porous liquids. At the molecular scale, bulky substituents and groups in the solvent molecule help to prevent the penetration of solvent into particles pores. In this study, a [4 + 6] imine-based porous organic cage (POC) was used as solid, and after evaluation of all the solvents through the models, 12 solvents were selected for experimental determinations, obtaining the best solvent for the fabrication of the porous liquid. The same approach was extrapolated to other imine-based POC. Even though the results obtained are satisfying, the authors conclude that their models would benefit by including other properties such as viscosity, as this is one important bulk property that may help the solvent to remain outside the cavities. R. J. Kearsey et al. (Kearsey et al. 2019) have proposed a high-throughput robotized workflow for the synthesis, characterization and testing of organic cages dissolved in solvents for the fabrication of type II porous liquids for the adsorption of xenon and methane, which allowed efficient choice and tests of a number of cage-solvents pairs.

Solution-based processing is omnipresent in industry. Porous liquids are a promising strategy to improve the processability of some solids that are otherwise difficult or impossible to disperse in a liquid and process as in the case of MOFs, COFs and so on. To face the major issues concerning processability of such solids and the possible solutions that include solvents, different studies have been presented. In the study developed by C. J. F. Oliveira et al. (Oliveira et al. 2020) where thin films of COFs were formed by their exfoliation through sonication in different solvents with different properties such as polarizability and hydrogen bonding ability, the Hansen parameters were taken into consideration as a guide for the selection of solvents for the preparation of the thin films. Sonication in the COF – solvent medium provoked bubbles formation, a linear relation was obtained between the ratio of these bubbles, interpreted as the resistance for the expansion of the bubble, and the polar Hansen parameter of the solvent. The roughness of the thin film was found dependent on this resistance as well. Control of Hansen parameters of the solvent medium may thus give a knob to control the thin film properties in these materials.

An example of this is the strategy developed by A. Knebel et al. (Knebel et al. 2020) to fabricate mixed matrix membranes formed by MOFs dispersed in a polymeric matrix though the formation of a porous liquid. For that, the outer surface of ZIF-67 nanoparticles was functionalized with N-heterocyclic carbenes. The dispersion of these particles was easily achieved in a solvent that afforded a porous liquid, and its properties of adsorption of CO_2, propylene and propane were tested. The porous liquids obtained were used for the fabrication of mixed matrix membranes by blending these liquids with a polymer matrix. The performances of the obtained membranes during adsorption and separation of propylene and propane were measured. The obtained characteristics of adsorption and separation of gases of the porous liquid obtained before membrane formation were maintained in the membrane. This study represents a step towards the use of porous liquids as a strategy to improve the processability of highly performant solids.

Physical priming is a technique consisting of the deposition of a small quantity of the polymer used as a continuous phase onto the surface of the particles used as filler to enhance the interactions by promoting interfacial adhesion between the particles and the polymer to improve the performance of the materials obtained, for instance, of membranes. In this method, a non-solvent is used so that the polymer adheres to the surface of the particles and avoids conglomeration, where primed particles adhere and form clusters. The strategy of physical priming is depicted in Figure 11 (Qian et al. 2020).

Z. Deng et al. (Deng et al. 2020) have presented a study concerning a Type II porous liquid formed by a metal organic cage in a solvent. The solvent was carefully chosen to ensure it remains out of the cage. After preparation, the porous liquid was encapsulated in graphene oxide nanoslits to form a porous liquid membrane. This membrane was tested for gas permeability, in particular to CO_2, H_2 and N_2. The authors found that the existence of pores in the membranes increased permeability to the cited gases. Interestingly, Type II porous liquids may present Henry and Langmuir adsorption modes together, increasing the capacity of adsorption during gas separation processes.

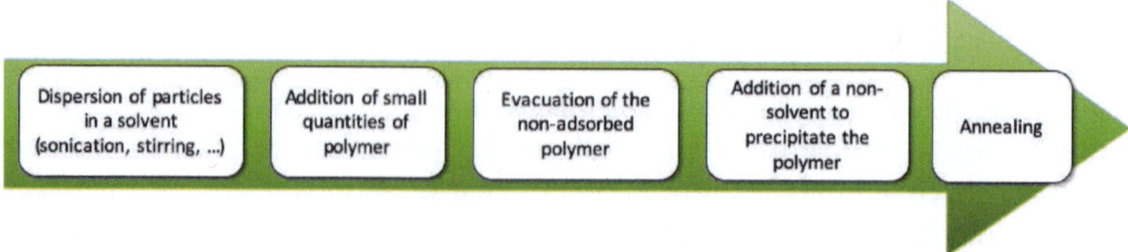

Figure 11. Physical priming steps to improve particle – polymer compatibility. Adapted from ref. (Qian et al. 2020).

Recently, I. Borne et al. (Borne et al. 2022) developed a study of a hypothetical gas absorption process by using a Type II porous liquid including CC13, which is a porous organic cage soluble in a variety of organic solvents. The analysis was performed by the McCabe-Thiele approach by considering an absorption and a desorption tower. The feeding stream contained 60% of CH_4 and 40% of CO_2. The selectivity towards CO_2 was evaluated together with tower sizes, yield of adsorption and solvent needs. The efficiency, number of equilibrium stages, energy required to regenerate the solvent and finally, the cost of the adsorption were used to evaluate the process and to compare it with a process using physical adsorption. Three desorption cases were suggested: pressure swing, temperature swing and pressure and temperature swing. To conclude, a decrease in the size of the absorption tower was achieved, and compared to the physical adsorption process using liquids. The use of the porous liquid allowed an increase in the gas capacity adsorption. However, deeper study is needed in order to improve selectivity.

The growing interest in developing porous liquids is justified for different applications. However, finding solid/liquid pairs suitable to form a porous liquid is not straightforward, and up to now the selections have been performed through trial and error. A recent contribution from the group of R. L. Greenaway and A. I. Cooper (Kearsey et al. 2019) proposed the use of a high-throughput robotic workflow including the synthesis, characterization and property testing of highly-soluble, vertex-disordered porous organic cages dissolved in a range of solvents excluded from the cavities of the organic cages. For that, an automatized high-throughput analysis of the solubility properties was performed, then the properties of the porous liquids were investigated, in particular the methane uptake. The porous liquids thus created presented an increased CH_4 uptake compared to the neat parent solvent.

Following with the idea of accelerating the prediction of solvent/solids pairs for the creation of porous liquids and of their properties, the group of R. P. Lively and D. S. Scholl has proposed using COSMO-RS calculations to accelerate the solvent selection for Type II porous liquids (Chang et al. 2022). This strategy would allow the selection of solvents from a library of more than 11,000 solvents available. This method has been tested for two molecular cages CC13 and TG-TFB-CHED, and even if experimental and predicted properties are not in perfect agreement, the method proposed can be used to find new Type II porous liquids with good accuracy.

Porous liquids represent a fascinating opportunity to perform Solvent Engineering for applications ranging from gases adsorption to separation. Solvent Engineering will be needed in this nascent field, as it is the role and the properties of the solvent that condition the obtaining of a porous liquid and the performance of the porous liquids once obtained.

CHAPTER 3
Engineering Aspects of Solvent Engineering

The engineering aspects of Solvent Engineering concern mass, heat and momentum transport. The effects of Solvent Engineering on these aspects are well observable in supercritical fluids and gas-expanded liquids. Engineering aspects are important as their consideration allows the optimization of industrial chemical processes.

3.1 Mass Transfer and Solvent Engineering

Density and viscosity play a key role in mass and momentum transfer, as can be observed in different heat, mass and momentum dimensionless transfer numbers such as the Reynolds number (Re), which represents the ratio of inertial forces to viscous forces within a fluid in movement (Equation 7)

$$\mathrm{Re} = \frac{\rho u L}{\mu} = \frac{u L}{v} \qquad \text{Equation 7}$$

where ρ is the density of the fluid in kg/m^3, u is the velocity of the fluid in m/s, L is a characteristic linear dimension in m, μ is the dynamic viscosity of the fluid in Pa.s, kg.m.s or N.s/m^2, and v is the kinematic viscosity in m^2/s.

The Schmidt number (Sc) represents the ratio of momentum diffusivity and mass diffusivity (Equation 8)

$$Sc = \frac{v}{D} = \frac{\mu}{\rho D} \qquad \text{Equation 8}$$

where D is the mass diffusivity in m^2/s.

The heat transfer analog of the Schmidt number is the Prandtl number (Pr), which represents the ratio of momentum diffusivity to thermal diffusivity (Equation 9)

$$Pr = \frac{v}{\alpha} = \frac{\mu/\rho}{\kappa/C_p\rho} = \frac{C_p\mu}{\kappa} \qquad \text{Equation 9}$$

where α is the thermal diffusivity in m²/s, κ is the thermal conductivity in W/m.K and C_p is the specific heat in J/kg.K.

The Lewis number (Le) represents the ratio of thermal diffusivity to mass diffusivity (Equation 10) and is used to characterize fluid flows where heat and mass transfer occurs simultaneously.

$$Le = \frac{\alpha}{D} = \frac{\kappa}{\rho D C_p} = \frac{Sc}{Pr} \qquad \text{Equation 10}$$

Density and viscosity play a determining role in particular applications and systems, such as determining the shape of bubbles or drops surrounded by a fluid. In this case, the Bond number (Bo), sometimes called the Eötvös number (Eo) (Equation 11) represents the extent of gravitational forces compared to surface tension forces

$$Bo = Eo = \frac{\Delta\rho g L^2}{\gamma} \qquad \text{Equation 11}$$

where g is the gravitational acceleration in m/s², L is the characteristic length in m, that can be for example the radii of curvature of a drop, and γ is the interfacial tension in N/m.

On the other hand, low viscosities, as observed in supercritical fluids or in gas-expanded liquids, lead to higher diffusivity, which enhances mass transfer in, for instance, separations and chemical reactions, by increasing the kinetics of the process. This is reflected by the Stokes-Einstein expression for the self-diffusion coefficient (Equation 12)

$$D_i = \frac{\kappa B^T}{6\pi\eta r} \qquad \text{Equation 12}$$

where κ_B is Boltzmann's constant, T is the temperature of the system, η is the dynamic viscosity and r is the radius of the particle, represented as a sphere. Concerning binary diffusion, which means the diffusion of a compound in a solvent media, it can be as well described by Equation 12, by fixing η as the viscosity of the solvent and r as the molecular radius of the solute (Kraft and Vogel 2017).

Viscosity is also important when talking about pressure drops along a pipe as is reflected in the Hagen–Poiseuille equation (Equation 13)

$$\Delta P = \frac{8\mu L Q}{\pi R^4}$$

Equation 13

where ΔP is the pressure difference between the two ends in Pa, L is the length of pipe in m, Q is the volumetric flow rate on m³/s and R is the pipe radius in m. It is easy to observe that engineering the viscosity of a fluid along a pipe would allow a decrease of pressure drop which can be traduced in a less energy consuming transport.

If one examines chemical reactions, density and viscosity of the solvent plays a key role. For instance, the dimensionless Damköhler number (Da) has been used extensively to classify a system according to the balance between transport and reaction, and is given as in Equation 14

$$Da = \frac{\tau_T}{\tau_R}$$

Equation 14

where τ_T is the transport (or residence) timescale and τ_R is the reaction timescale (Oldham et al. 2013).

It is important to note here that a lot of simplifications of the transport equations assume an incompressible fluid, which, in the case of supercritical fluids and of gas-expanded liquids, may not be pertinent. One example of this is the second viscosity. In fact, shear viscosity plays a key role when fluid dynamics is studied; it can be at the origin of instabilities that can lead to flow turbulence (Cramer 2012). In common applications, the viscous stresses are mostly well described by the shear viscosity (η). However, a complete description needs the description of the bulk viscosity, sometimes called volume or dilatational viscosity (κ) (Jaeger et al. 2018), meaning that isotropic dilatations of a volume element of fluid contribute to the viscous stresses of the fluid. κ is defined as in Equation 15,

$$\kappa = \lambda + \frac{2}{3}\eta$$

Equation 15

where $\kappa = \kappa(P, T)$, $\lambda = \lambda(P, T)$ and $\eta = \eta(P, T)$ are the bulk, second and shear viscosities respectively, and P and T are the absolute pressure and temperature. Stokes' assumptions stated that $\lambda = -\frac{2}{3}\eta$; and hence $\kappa = 0$, which is often used as simplification. However, in certain processes, taking the second viscosity into account is necessary, for example in nanochannel flows where large variations in the pressure and fluid density are known to occur, in rapid expansion of a supercritical solution (RESS), which is a process to fabricate solid particles from a solution, and in processes involving gas expanded liquids given their high compressibility, etc. It is noteworthy that bulk viscosity for CO_2 has been studied and found to be between 2000 (Tisza 1942) to 3000 (Emanuel 1992) times its shear viscosity. These abnormally high values of bulk viscosity of CO_2 can be explained by the intramolecular relaxation times that are in the µs range and the low vibrational wave numbers (< 1000 cm^{-1}) for this molecule. In these cases, the dilute gas contribution due to the relaxation of internal degrees of freedom is seen to be significant (Jaeger et al. 2018). Other fluids have been studied such as air, methane, ethylene, and hydrogen, among others. However, no data concerning the second or bulk viscosity of a gas-expanded liquid has been found in literature and even measurements for gases are scarce. This lack of experimental data may arise from the difficulty of measurement of this property, that needs measurements of acoustic absorption, relaxation times or shock waves (Cramer 2012). It is however important to say that ignoring the contribution of λ in applications such as supersonic combustion, Rankine cycle power systems using working fluids such as steam, can result in important deviations. Figure 12 shows a thermodynamic P-V diagram with the anomalous region of dense flows.

Dense gases with large specific heats, known as Bethe-Zel'dovich-Thompson (BZT) fluids, possess regions of negative non-linearity near the saturation-vapor line at pressures and temperatures approaching their critical point values. Regions of a flow field that exhibit compressible thermodynamic parameter (Γ) $\Gamma < 0$ are said to possess negative non-linearity, while those with $\Gamma > 0$ have positive non-linearity. In flows having $\Gamma < 0$, conventional compression shock waves and expansion fans are not possible. The only waves that are physically possible are expansion shock waves and compression fans, which follows from the second law of thermodynamics (Graves and Argrow 1999). Concerning CO_2, which, as presented, is one of the most used fluids in supercritical form or in gas-expanded liquids, Gonzalez and

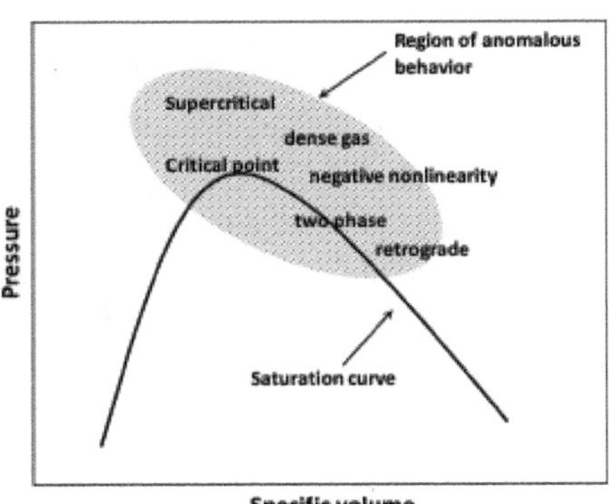

Figure 12. Dense gas region of anomalous behavior, adapted from reference (Graves and Argrow 1999).

Emanuel (Gonzalez and Emanuel 1993) studied the effect of bulk viscosity in Couette flow of CO_2 with porous walls and bulk viscosity effects related to the boundary layer of a high-speed flow of CO_2 over a flat plate. It is to be noted that CO_2 presents a ratio β/μ large, around 2000, which needs to be taken into account for certain applications, keeping in mind that bulk viscosities for dense fluids are poorly understood.

Calculation of the shear and bulk viscosities can be performed by using the Green-Kubo expression by using atomistic and coarse-grained force fields and from Einstein relations (Jaeger et al. 2018), SAFT-γ coarse-grained models provide a computationally-efficient alternative to classical atomistic force fields with a good level of accuracy for shear viscosity calculations (Jaeger et al. 2018). It is relevant that κ/η has been found to be around 3 for water and n-decane but can reach up to 200–2300 for liquid and gas phase CO_2 respectively. Sensitive applications concerning CO_2, including supercritical CO_2 applications or CO_2-expanded liquids usages need to take into account this specificity of CO_2 and of other compressible fluids, including gases, for which, unfortunately, literature data is extremely scarce. The study performed by Z. Xu et al. (Xu et al. 2019) shows the effect of solvent properties, among other characteristics, in the design of mass transfer equipment such as a packed column for CO_2 absorption by a solvent in countercurrent flow. It must be remembered that an important value of solvent viscosity can lead to a significant reduction in the mass transfer coefficient and diffusivity, as can be seen in Equation 15. High solvent viscosity can lead to slow diffusion of gas or of a solute through a gas-liquid interface or across the bulk phase to the surface of a catalyst, which can modify the local hydrodynamics by generating a more stabilized liquid film on the interface or on the surface of the catalyst. As a result mass transfer between phases can be significantly reduced (Song et al. 2018; Xu et al. 2019).

One example of the control of physico-chemical properties of a solvent by simply mixing with another solvent was exemplified by the work of V. S. Protsenko et al. (Protsenko et al. 2015), where the physico-chemical properties of mixtures of choline chloride, ethylene glycol, nickel chloride and water were studied. In this study, a decrease in density, viscosity and surface tension was observed with an increasing quantity of water, while conductivity increased with water content. The increase of temperature had an important effect too by decreasing density, surface tension, viscosity, while conductivity increased with an augmentation of temperature. This study represents an easy method to optimize the mass transfer properties of a solvent for a desired application, where transfer properties can be crucial, such as diffusion of bulky substrates in a chemical reaction.

3.2 Thermo-physical Properties and Solvent Engineering

Thermodynamics is the study of systems involving energy in the form of heat and work. The first law of thermodynamics expresses the relationship between internal energy to heat added to or extracted from the system, together with the work performed by it. It reflects the conservation of energy, and in the case where changes in potential and kinetic energies are negligible, it takes the form of

$$\Delta U = Q - W \qquad\qquad \text{Equation 16}$$

where U is the thermal energy, Q is the heat input to the system and W is the work done by the system. In an isothermal process, the temperature stays constant, so the pressure and temperature are inversely proportional to one another. If pressure and volume are represented in a graph for an ideal gas, an isothermal process looks like in Figure 13.

In an adiabatic process, no heat is added or removed from the system and Equation 17 takes the form:

$$\Delta U = -W \qquad\qquad \text{Equation 17}$$

By using the first law, the propriety enthalpy (H) can be defined as:

$$H = U + PV \qquad\qquad \text{Equation 18}$$

H is a function of the state of the system and has units of Joules.

To know the changes in temperature that a given amount of heat produces in a substance, a constant (C) can be defined:

$$Q = C\Delta T \qquad\qquad \text{Equation 19}$$

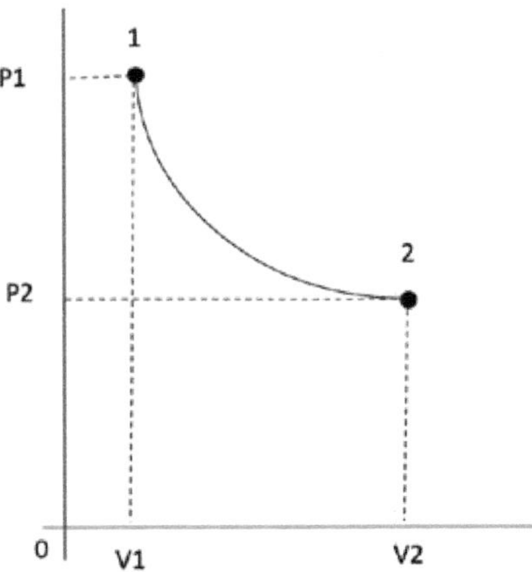

Figure 13. Diagram Pressure vs Volume for an isothermal process in an ideal gas.

If the process is at constant pressure, the constant is called the specific heat at constant pressure (C_p), if it is determined at constant volume it is called the specific heat at constant volume (C_v), and the definitions are:

$$C_p = \left(\frac{\partial H}{\partial T} \right)_P$$

Equation 20

and

$$C_v = \left(\frac{\partial U}{\partial T} \right)_V$$

Equation 21

The coefficient of isothermal compressibility (κ_T) of a pure fluid is given by Equation 22, and represents the fractional change of fluid volume per unit change in pressure. This property decreases with pressure. However, the isothermal compressibility of a pure compound presents a maximum value around the critical point. This behavior has been explained by the changes in density observed near the critical point.

$$\kappa_T = -\frac{1}{V} \left(\frac{\partial V}{\partial P} \right)_T$$

Equation 22

The coefficient of thermal expansion is defined as:

$$\alpha = \frac{1}{V} \left(\frac{\partial V}{\partial T} \right)_P$$

Equation 23

and the coefficient of isobaric thermal expansion as:

$$\beta = \frac{1}{V} \left(\frac{\partial V}{\partial T} \right)_P$$

Equation 24

Clapeyron relationships indicate how the thermal capacities depend on V and P, which allow us to obtain these properties from an equation of state, C_p has a maximum value at the critical point for water, CO_2, and other fluids (Pioro 2011).

$$\left(\frac{\partial C_v}{\partial V} \right)_T = T \left(\frac{\partial^2 P}{\partial T^2} \right)_V$$

Equation 25

$$\left(\frac{\partial C_p}{\partial P} \right)_T = -T \left(\frac{\partial^2 V}{\partial T^2} \right)_P$$

Equation 26

C_p and C_v can be determined experimentally, as it can be observed in literature data for different systems such as pure CO_2 (Colina et al. 2003), methanol-CO_2 (Ishmael et al. 2017), ethanol-CO_2 (Boulton and Stein 1993) and so on. It is, however, pertinent to note the lack of exhaustive experimental data for systems concerning hydrogen and supercritical fluids. Calculations from Equations of State, or from Monte Carlo-based models can be performed when data are needed on these and on other systems.

The values of thermo-physical properties of solvents have a significant effect on the design of physico-chemical processing, reaction units and heat transfer devices, as they directly influence the design parameters and performance of equipment such as heat exchangers, distillation columns, chemical reactors (de Castro et al. 2012). Solvent Engineering can be applied to different solvent systems in order to obtain the best heat exchange properties and to adapt them to the chemical process developed.

3.3 Thermodynamics and Solvent Engineering

Thermodynamics plays a key role during solubilization and solvation. For instance, the activity coefficient of a compound (γ) is a parameter used to take into consideration the deviations from ideal behavior of a mixture. If $\gamma = 1$, the solution is an ideal solution, if $\gamma < 1$ or $\gamma > 1$, the solution will present respectively negative or positive deviations from Raoult's law. Activity coefficients greater than 1 indicate that thermodynamic forces are not favorable to the solubilization of the compound. Activity coefficients can be calculated theoretically by using models such as UNIQUAC, NRTL, UNIFAC and MOSCED, among others, for non-electrolytes; these models often need experimentally fitted parameters.

J. W. Lorimer (Lorimer 1993) has deduced an equation to describe the solubility of a solute that is completely ionized and that dissolves in a mixed solvent. This equation takes into consideration the solubilities of the solute in the pure solvents individually, the properties of the mixture of solvents and other parameters that present molecular significance and that have been deduced from classical thermodynamics and exact statistical thermodynamic theory of liquids. This model contains three parameters: excess solubility, excess activity coefficient and excess chemical potential. The excess solubility was calculated from classical thermodynamics and experimental solubility data; the excess activity coefficient was calculated from Pitzer's equation for activity equation, provided that all the parameters are known; the last term of the equation was calculated from the theory of Kirkwood and Buff.

The SAFT (Statistical Associating Fluid Theory) family of Equations of State (EoS) is a family of theoretically based engineering EoS that has been derived from the model proposed by Chapman et al. (Chapman et al. 1989), and which is based on the perturbation theory of Wertheim (Wertheim 1984). For instance, one of these EoS is the PC-SAFT (perturbed chain-SAFT), introduced by Gross and Sadowski (Gross, J.; Sadowski, G. Ind. Eng. Chem.

Res. 2001, 40, 1244.) which is based on a coarse-grained model that represents the molecules as a chain of spherical segments interacting with other chains. In the PC-SAFT model, non-polar, non-associating molecules are characterized by three parameters of the pure components: the number of segments per chain molecule, the segment diameter and the dispersion energy. For polar molecules it is necessary to include the dipole moment and the quadrupole moment. Two additional pure-component parameters are used for the associating interactions: the association energy and the effective associative volume. Prediction of the phases equilibrium was performed in the cited study by using PC-SAFT as well, with good agreement of experimental data.

O. Odele and S. Macchietto (Odele and Macchietto 1993) have proposed to use a mixed-integer non-linear programming (MILP) method coupled to a group contribution-based molecular design to the solvent design for liquid-liquid extractions and gas absorption. In this study, physical and equilibrium properties of the molecules were obtained by a combination of group contribution and empirical correlations: for instance, activity coefficients by UNIFAC model, vapor pressure by Riedel model, Fugacity coefficients by UNIWAALS model, and so on. J. Noroozi and W. R. Smith (Noroozi 2021) proposed a hybrid approach between classical molecular dynamics simulations with the General Amber Force Field and semi-empirical AM1-BCC partial charges assignment method for the amine molecules and their protonated and carbamate forms (GAFF/AM1-BCC) in the liquid phase, together with quantum chemical ideal-gas calculations for the gas phase, and applied this approach to the prediction of thermodynamic properties of amine species of CO_2-loaded aqueous solutions of alkanolamines for the design of carbon capture solvents. Their results are within an absolute error of ≈ 4 kJ in the reaction free energy values and less than 0.72 pK_a units by studying a set of 77 amines at different temperatures.

In an effort to make clear the relationship between solvation properties and other thermodynamic properties that are usually calculated in chemical engineering, S.-T. Lin et al. (Lin et al. 2007) have developed equations that take into consideration classical thermodynamics that have been represented by the compressibility factor, statistical thermodynamics, represented by the Helmholtz free energy of conformation, and solvation thermodynamics, represented by the total solvation free energy. They have derived the equations for thermodynamic properties used during phase equilibrium calculations in terms of molecular solvation free energy. These authors have compared their results with the van der Waals EoS and showed that total solvation free energy, configuration Helmholtz free energy and compressibility factor can be related between them, and a complete treatment can be used during phase equilibrium calculations.

The choice of judicious co-solvents for a given reaction is often performed empirically and by trial and error experiments. A. Wangler et al. (Wangler et al. 2019) have proposed a method to predict the influence of co-solvents on Michaelis constants for enzyme-catalyzed reactions through a thermodynamic activity-based approach. The approach proposed by these authors is the use of electrolyte perturbed-chain statistical associating fluid theory (ePC-SAFT), which is claimed to be suitable to predict activity coefficients for bio reactions. This approach has been found valid for the prediction of initial reaction rates and future work is needed to predict the entire progress-time evolution. The authors also suggest using molecular dynamics simulation techniques to take into consideration the inhibition effects of the solvent on the enzyme.

In their work that presents the use of Computer Aided Molecular Design (CAMD) method to support the synthesis and selection of molecules in process systems and to guide in experimental works the choice between several molecules, A. Papadopoulos et al. (Papadopoulos et al. 2016) have developed an approach for the choice of solvents for the chemisorption of CO_2 where thermodynamics, reactivity and sustainability were taken into consideration to screen the candidate molecules. To predict the non-equilibrium behavior and the phase equilibrium of the mixture, the group-contribution statistical associating fluid theory for square well potentials (SAFT-γ *SW*) was used. Concerning the thermodynamic treatment, relative energy difference (RED) compounds, together with vapor pressure, liquid heat capacity (Cp), density, surface tension, viscosity, melting and boiling points were used as thermodynamic parameters. This method seems to be effective for the design and choice of candidate solvents for specific applications. Comparison with experiment results would be needed to validate the obtained choices.

The challenge of solvent selection for the absorption of CO_2 is not new, and a good number of studies and works have been published dealing with it. Among the works addressing this challenge, including a thermodynamic study, is the one from M. Stavrou et al. (Stavrou et al. 2014), that proposed a continuous molecular targeting – CAMD (CoMT - CAMD) method, with thermodynamic properties prediction by using the PC-SAFT model.

CAMD approaches have been used for the design of solvents for reactions by M. Folić et al. (Folić et al. 2008) with the aim to maximize product formation. A multi-parameter solvatochromic equation (presented in Equation 27) was used that correlates the logarithm of the reaction rate constant with properties of the solvents such as solvatochromic parameters (A, B and S), a polarizability term (δ), and the Hildebrand solubility parameter (δ_H).

$$\log k = \log k_0 + sS + d\delta + aA + bB + \frac{h\delta_H}{100} \qquad \text{Equation 27}$$

where $\log k_0$, s, d, a, b and h are constants that are specific to the studied reaction and quantify the dependence of the rate constant on the solvent. A is the hydrogen bonding acidity, B is the hydrogen bonding basicity, and S the dipolarity/dipolarizability. Two types of model reactions were taken into consideration: competing reactions and consecutive reactions. One advantage of this approach is that its use and development do not require deep understanding of the mechanism of the reaction or of the effects of the solvents at the electronic level.

N. P. Mendis et al. (Mendis et al. 2022) have developed a study concerning the simultaneous solvent selection and solvent design for a chemical process involving reaction, extraction and crystallization of a compound, with the possibility of solvent recycling. The proposed equilibrium-based framework has been applied to the continuous synthesis and purification of dalfampridine, which is a compound used for the treatment of multiple sclerosis. During this study, the electrolyte perturbed-chain statistical association fluid theory (ePC-SAFT) EoS was used to select solvents and simultaneously optimize the operating conditions of the continuous process together with a mixed-integer nonlinear programming problem (MINLP) and a local nonlinear programming problem (NLPs) approach. This study was a continuation of a previous work for continuous crystallization process with solvent recycling (Mendis et al. 2020; Wang and Lakerveld 2018a, 2018b; Wang et al. 2020). The strategy proposed in this study allowed the consideration of

1) liquid-liquid, solid-liquid and vapor-liquid equilibria simultaneously,
2) the impact of ions that may be present in solution,
3) the use of multiple crystallization steps and
4) the estimation of pure component data.

A sensitivity analysis of costs, including solvents, heating and cooling was performed, showing that the optimal solvent types remained constant even with changing the costs factors.

Solvent Exchange thermodynamics to induce folding or unfolding of macromolecules, in particular proteins, was developed by J. A. Schellman (Schellman 1994). The model presented was developed by extending a previous one that concerned free energy and preferential interactions (Schellman 1987, 1990). The last extended model included thermal properties treatment such as enthalpy, entropy and heat capacity. The model, however, presents some limitations such as the assumption that exchange sites are independent, and exchange is carried out in a one-to-one fashion for site to solvent molecules.

3.4 Heat Transfer and Solvent Engineering

Heat is transferred by conduction, convection and radiation. In conduction, the substance does not move but transfer of energy arrives through contact within a substance or between substances in contact. In convection, the motion of the fluid carries heat from one place to another, while radiation involves the absorption or giving off of electromagnetic waves. Conduction and convection rely on differences of temperature. Most of the time, heat transfer happens through two of the three mentioned mechanisms, even if one may be predominant.

Fourier's law dictates the conduction of heat, and stipulates that heat transfer through a material is proportional to the negative gradient of temperature. This law can be used in integral form to calculate the amount of energy flowing into or out of a body and the differential form to calculate local flow rates or fluxes of energy. In its differential form, it is expressed as:

$$q = -\kappa \nabla T \qquad \text{Equation 28}$$

where q is the local thermal flux density, κ is the thermal conductivity and ∇T is the temperature gradient. It is to be remembered that ∇T varies with temperature and if the material is not isotropic, ∇T may vary with orientation, then it will be represented by a second-order tensor. When taking into consideration only the x direction, Equation 28 takes the form:

$$q_x = -\kappa \frac{dT}{dx} \qquad \text{Equation 29}$$

The integral form of the Fourier's law is:

$$\frac{\partial Q}{\partial t} = -\kappa \oiint_s \nabla T \, dS \qquad \text{Equation 30}$$

where $\dfrac{\partial Q}{\partial t}$ is the amount of heat transferred par unit of time and dS is a surface area element. If this equation is integrated, and for a homogeneous material of 1-D geometry, between two surfaces at a constant T, the flow rate gives:

$$\frac{Q}{\Delta T} = -\kappa A \frac{\Delta T}{\Delta x} \qquad \text{Equation 31}$$

where A is the cross-sectional surface area, ΔT is the difference in temperature between the surfaces, and Δx is the distance between the surfaces.

When talking about heat transfer in heterogeneous catalytic hydrogenation reactions in sub or supercritical conditions, and given the high exothermicity of the reaction, it is interesting to understand whether convective or conductive heat transfer is predominant in order to evaluate the 'quality' of the heat transfer. The definition of a dimensionless number, the Nusselt (Nu), as the ratio between convective to conductive heat transfers, is useful:

$$Nu = \frac{hL}{\kappa} \qquad \text{Equation 32}$$

where h is the convective heat transfer coefficient and L is a characteristic length.

The Prandtl number (Pr) is the ratio of momentum diffusivity to thermal diffusivity and is calculated as:

$$Pr = \frac{C_p \mu}{\rho} \qquad \text{Equation 33}$$

where μ is the dynamic viscosity of the fluid.

Some values of thermo-physical properties for some solvents currently used are shown in Table 1.

Besides thermal conductivity, specific heat capacity of solvents is of great importance for practical applications, such as for temperature-sensitive chemical reactions or for the development of energy-effective chemical processes. Knowledge of this property is then essential for the determination of heat transfer properties such as the calculation of enthalpies, and flow properties used in chemical processes design and simulation. Solvent Engineering can be applied to heat transfer properties in order to obtain the best conditions to perform the optimization of the process carried out in the engineered solvent.

For instance, heat transfer during the flow of supercritical fluids in pipes has been fairly extensively studied in recent years paying particular attention to the behavior of transport and physico-chemical properties at the critical point (Song et al. 2008; Zhang et al. 2018; Lopes 2017; Huai and Koyama 2007). Some studies have been published concerning hydrogenation in supercritical phases (Härröd et al. 2001), such as the work performed by H. Häring et al. (Häring et al. 2017) concerning the development of a model for the hydrogenation of citral in a fixed bed reactor and a Pd/alumina catalyst. Mass and heat transfer was taken into consideration together with other variables relevant for the process such as particle diameter, H_2:citral ratio and CO_2:citral ration variations, among others. The AspenTM software was successfully used for the simulation of the reactor.

On the other hand, ionic liquids have been proposed in applications where specific thermo-physical and chemical properties are needed. Ionic liquids are salts that are liquid at or near room temperature. As ionic liquids may be seen as other solvents, their thermo-physical properties need to be obtained. There is good potential to modulate these and other properties of ionic liquids by changes in the ion and in the cation in the ionic liquid structure. Examples of

Table 1. Thermophysical properties for some solvents currently used.

Solvent	κ (W m^{-1} K^{-1})	η (mPa·s)	C_p (J kg^{-1} K^{-1})	ρ (kg m^{-3})	Ref.
Water	0.631	0.653	4 179	992	(de Castro et al 2012)
Ethylene glycol	0.256	10.37	2 520	1 100	(de Castro et al 2012)
[C$_4$mim][NTf$_2$]	0.116	28.50	1 372	1 423	(de Castro et al 2012)
[C$_4$mim][EtSO$_4$]	0.178	50	1 615	1 226	(de Castro et al 2012)
Glycerol (25°C)	0.292	700	2 409	1 250	(Singh et al 2018)
[ChCl][glycerol] (molar ratio 1:2) (25°C)	0.21	200	-	1 185	(Singh et al 2018)
Acetone (25°C)	0.161	0.308	2 160	784.2	(Malhotra and Woolf 1991)
Ethanol (25°C)	0.169	1.082	2439.77	785.46	(González et al. 2007)
Methanol (25°C)	0.200	0.545	1928.35	786.9	(González et al. 2007)
Dimethylformamide (25°C)	0.183	0.827	2 070	944.6	(Bernal-García et al. 2008; Hao et al. 2019; Rodríguez-Laguna et al. 2018)
Ethyl acetate (25°C)	0.143	0.426		894.43	(González et al. 2007)

Table 2. Some thermo-physical properties of some ionic liquids.

Property	Ionic liquid		
	[EMIM][BF$_4$]	**[BMIM][BF$_4$]**	**[DMPI]IM**
Density at 60°C (g cm^{-3})	1.2526 ± 0.0001	1.1747 ± 0.0001	1.4205 ± 0.0001
Melting point (°C)	14.42 ± 0.33	−87.38 ± 2.43	11.29 ± 0.01
Heat capacity at 100°C (J g^{-1} K^{-1})	1.28	1.66 ± 0.08	1.2 ± 0.05
Thermal conductivity (298 K) (W m^{-1} K^{-1})	0.200 ± 0.003	0.186 ± 0.001	0.131 ± 0.001
Viscosity (298 K) (mPas)	36.07 ± 0.17	119.78 ± 1.28	90.05 ± 0.51

the dramatic incidence of different cations and anions in the thermo-physical properties of ionic liquids can be observed in Table 2 (Valkenburg et al. 2005).

3.5 Solvation Properties and Solvent Engineering

Using judiciously engineered solvents allows us to overcome problems such as poor solubility of reactants. Solubility of gaseous reactants such as H$_2$ or N$_2$ is a well-known issue in catalysis.

One elegant example of the engineering of solvents to resolve these problems is the use of gas-expanded liquids to improve the solubility of gaseous reactants such as H$_2$ by taking advantage of the complete miscibility of CO$_2$ with these gases, as has been shown during the continuous hydrogenation of limonene performed by E. Bogel-Lukasik et al. (Bogel-Łukasik et al. 2008) by using a carbon-supported palladium catalyst.

CHAPTER 4
Fields of Application of Solvent Engineering

4.1 Green Chemistry and Engineering

Among the principles of Green Chemistry and Green Engineering stated by Paul A. Anastas (Anastas and Eghbali 2010; Morgenstern, D. A.; LeLacheur, R. M.; Morita, D. K.; Gross, M. F.; Burk, M. J.; Tumas 1996; Mulvihill et al. 2011), three guidelines are closely linked to the use of solvents in a chemical process:

1) synthetic methods should be designed to use substances with little toxicity to human health and the environment
2) safer solvents need to be used, and
3) Design for energy-efficient processes.

In this context, Solvent Engineering has found an excellent field of application in Green Chemistry, as judiciously engineered solvents can serve to decrease the usage and even to replace traditional, highly toxic solvents currently used for extractions, separations, chemical reactions, and so on. Noteworthy here is the remarkable work of P. G. Jessop and M. Poliakoff in the fields of switchable solvents and of supercritical fluids respectively, both of them applied to the development of greener chemical processes. An excellent example of a fruitful collaboration between academic research and industry was developed between the Clean Technology Group at the University of Nottingham and Thomas Swan & Co. Ltd., a fine chemicals manufacturer (Licence et al. 2003), to perform chemical reactions in $scCO_2$, in particular to perform continuous hydrogenation using heterogeneous catalysts. As $scCO_2$ is completely miscible with H_2 and N_2, its use allowed an important concentration of these reactants in the reaction media, which is impossible to achieve by using liquid organic solvents. A plant was constructed under this principle in 2002.

The possibility to decrease the quantity of solvent used during a chemical process by using solvent engineering and in particular gas-expanded liquids has been shown by a few but concrete examples that can be found in literature. One of these examples was presented by E. Siougkrou et al. (Siougkrou et al. 2014) through the development of a methodology for the design of GXLs based on the performance of the GXL in a chemical process. This methodology proposes to choose the best GXL based on the overall economic performance of the conceptualized process. In this study, the addition of CO_2 to the organic solvent is compared to the use of pure organic solvent within the process. It is to be noted that the nature of the GXL regarding the organic solvent and the composition in CO_2 are key decisions that impact reactor volumes and energy needs of the process. In this study, questions such as the effect of GXL design on reaction rate and on the solubility of the reactants as a function of CO_2 composition are addressed through the use of empirical models that link the properties of the GXL to reaction rate constants and a predictive equation of state to take into account the effect of temperature on the system. A case study is proposed in this article which concerns the Diels-Alder reaction of anthracene and 4-phenyl-1,2,4-triazoline-3,5-dione (PTAD) to form the adduct 8,9,10,11-dibenzo-4-phenyl-2,4,6-triaza[5,2,2,0] tricyclo-undeca-8,10-diene-3,5-dione.

Interestingly, solvent effects on the reaction rate were incorporated by introducing an empirical model which correlates reaction rate constants, or other free-energy based properties such as solubility and octanol-water partition coefficients, with solvent properties (Ford et al. 2008). This is called a solvatochromic equation; the original form is shown in Equation 34:

$$\ln k_{r,i} = \ln k_{r,0} + s_r\, \pi_i^* + a_r\, \alpha_i + b_r\, \beta_i \qquad \text{Equation 34}$$

where $k_{r,i}$ is the rate constant for a reaction r in a solvent i, $k_{r,0}$ is the rate constant for reaction r in a reference solvent, π^*, α and β denotes the solvatochromic parameters of the solvent (polarity, hydrogen bond acidity and hydrogen bond basicity, respectively); s_r, a_r and b_r are coefficients that depend on the reaction of interest. These coefficients are generally regressed on experimental kinetic data in several solvents.

For the reaction considered, the kinetic data in CO_2-expanded acetonitrile was used to obtain the pseudo-first order rate constant as function of the properties of the GXL, and Equation 34 takes then the form:

$$\ln k_i\,(x_{CO_2}) = 1.9 - 2.62\pi_i^*\,(x_{CO_2}) - 4.68\alpha_i\,(x_{CO_2}) + 1.58\beta_i\,(x_{CO_2}) \quad \text{Equation 35}$$

Application of this equation to obtain the reaction rate constant (k_i) in three GXLs as a function of CO_2 mole fraction, is shown in Figure 14a. The total cost of the process was calculated by taking into account material

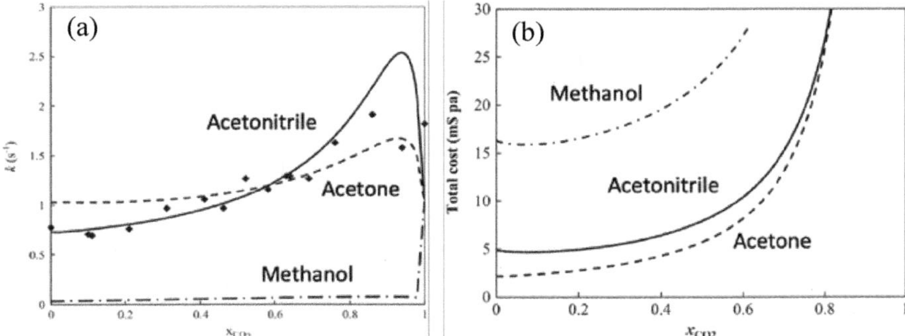

Figure 14. (a) Calculated pseudo-first order reaction rate constant, k_i, in mixed solvents, as a function of CO_2 mole fraction. (b) Calculated total process cost as a function of CO_2 mole fraction in the GXL for a single-pass conversion of 0.5. Taken from reference (Siougkrou et al. 2014) with permission of Elsevier.

balances, capital and operating costs, and by considering a process consisting of a CSTR, a separation unit and a compressor (Figure 14b).

It is to be noted from Figure 14 that the nature of the GXL has a net impact on reaction rate and on the total cost of the process. However, the trend is not the same for both variables, as in the case of acetone, where intermediate reaction rates and the most competitive costs for the process were obtained. Higher costs of the process were, however, linked to the use of a compressor and hence, higher amounts of CO_2 in the GXL increased the cost of the entire process. It is, however, important to say that the decrease of the use of organic solvent allowed by the use of GXLs needs to be taken into account from an environmental point of view for new processes. In the case presented here, economic and environmental indicators seem to encourage the use of up to 10–15 mol% of CO_2 in acetonitrile, which has a minimal repercussion on process costs and decreases solvent usage by 33%.

Another example of the usage of Solvent Engineering in Green Engineering is the use of a CXL in the work presented by Z. Xie and B. Subramaniam (Xie and Subramaniam 2014) concerning the 1-octene hydroformylation process. In this study, the Rh-catalyzed process was compared with the Co-catalyzed process in CO_2-expanded toluene. The process flow diagram of the simulated conventional process and that of the simulated CXL process is shown in Figure 15(a) and (b) respectively for a production of 150 kton/year. Aspen® HYSYS with the UNIQUAC package for relevant thermodynamic properties prediction and the Peng-Robinson equation of state for liquid-vapor phase modeling were used; mass and energy flows obtained from HYSYS simulations were used for economic and environmental analyses. Life Cycle Analysis (LCA) was performed by using a commercial software called GaBi® and the Economic Input-Output Cycle Assessment (EIO-LCA)

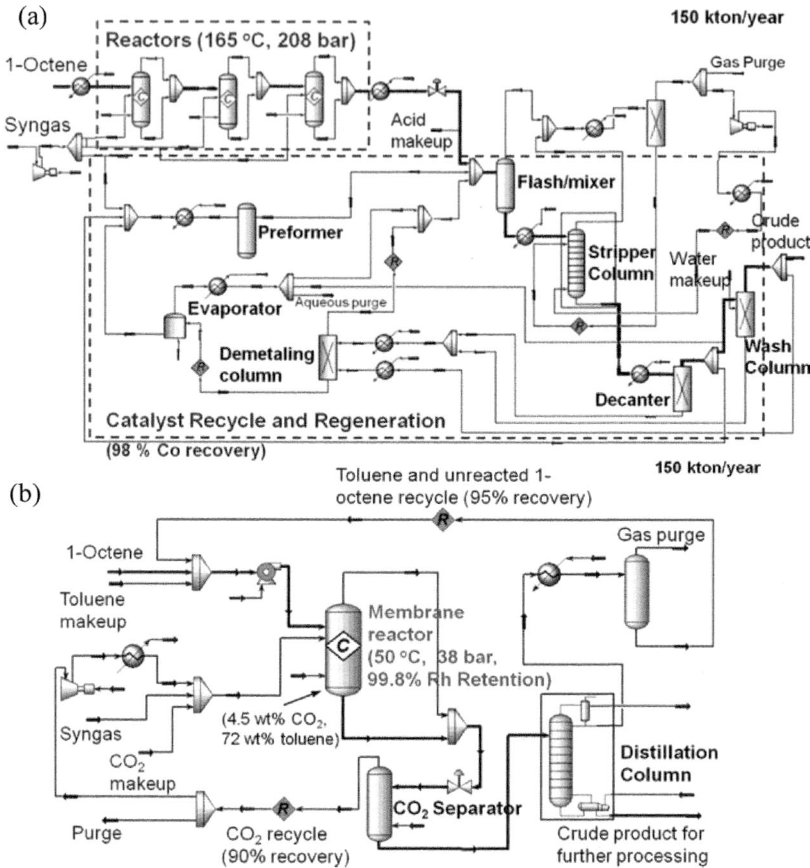

Figure 15. Process flow diagram of (a) the simulated conventional process and (b) the simulated CXL process. Reprinted with permission from (Xie and Subramaniam 2014). Copyright (2014) American Chemical Society.

tool. The results of this study show that the capital investment was 30% lower for the CXL process, in comparison to the conventional Co-catalyzed 1-octene hydroformylation process of similar capacity. The LCA results show that the CXL process is more environmentally friendlier than the conventional process, and that in general, the CXL process generates less emissions within the plant than the conventional process.

It is to be noted that in the context of Green Chemistry, Solvent Engineering can and should be associated with greener solvents. However, this association imposes new paradigms in research and in industrial applications as, for instance, greener solvents in general have lower boiling points, making difficult separation by distillation difficult. They may present poorer solvation properties than traditional solvents and less data is available concerning industrial or laboratory applications. In this context, a number of literature reviews and books have been published regarding the use of green solvents for sustainable chemistry and chemical processes (Chapter 1 Introduction to Solvents and Sustainable Chemistry 2017; Pollet et al. 2014; Schuur et al. 2019; Sheldon 2019; Welton 2015). In their recent review, Coby J. Clarke and coll. (Clarke et al. 2018) present a comprehensive overview of the most cited sustainable organic solvents proposed up to now such as ionic liquids, DES, supercritical fluids, switchable solvents, liquid polymers and renewable solvents. The authors painstakingly present the performance of each the solvents presented in the context of the chemical process that makes use of it, and consider as well the wider context of the process or system that uses the solvent, taking technical, economic and environmental parameters into consideration. One part of this review is devoted to switchable water, which is produced by the introduction of CO_2 into a mixture of water and an organic, water-miscible base, leading to the formation of ammonium bicarbonate salts. CO_2 in this cases triggers a change in ionic strength, decreasing the solubility of organic compounds, and increasing conductivity and viscosity of the solution (Mercer and Jessop 2010). Switchable ionic liquids, formed by an organic base and an alcohol can be envisaged to be used in gas capture and in separation processes. These solvent mixtures change between ionic and non-ionic form upon absorption of an acidic gas, which represents switchable solvent properties that are highly desirable in green chemistry and processing. One of the most commonly used amine is 1,8-diazabicycloundec-7-ene (DBU), which reacts with CO_2 in a molar ratio of 1:1 to form carbonylated ion pairs (Yao et al. 2017).

The application of green metrics to solvents has been discussed in the review by T. Welton (Welton 2015). An interesting observation pointed out in this review is that replacing a solvent by a greener one may decrease the efficacy and hence the sustainability of an entire process; it is then crucial to consider the sustainability of all the processes when suggesting the change of a solvent. For instance, solvent selection guides such as the one developed by Pfizer (Alfonsi et al. 2008) or Sanofi (Prat et al. 2013), among others, offer good suggestions for the replacement of the current solvent by less toxic, less ozone depleting solvents among other properties. However, these guides do not place the solvent replacement in the context of the entire process. For that, the development of a life cycle analysis of the new process is advisable.

Ortho-xylene has been proposed as greener solvent for the processing of organic photo-detectors through a solution process. The polymer PBDB-T and the non-fullerene acceptor CO_1-4F were both solubilized in this solvent. Excellent photo-sensing properties were obtained in the fabricated device, comparable with those obtained by using chlorinated solvents, stressing the point that green solvents can be used in high-performance innovative devices and that research concerning these solvents is still in its infancy (Huang et al. 2023).

Solubilization of polymers is a ubiquitous step in many industries and activities. Paints and coatings industries are the most solvent-consuming industries, consuming 46% of all the solvents consumed by different industries and manufacturing activities, as pointed out by A. Zhenova (Zhenova 2020). The replacement of toxic and polluting solvents of polymers in industry by non-toxic and environmentally-acceptable solvents or by using water-based formulations is a challenge that has been addressed, for instance, in projects funded by different agencies and private sectors. The PROVIDES project was funded with one million euros to study the use of deep eutectic solvents for lignocellulose, while the R2LiB project was funded with £2.5 million to study the use of green solvents for batteries recycling. Innovative solvents such as Cyrene have been developed during a collaboration and production is now being scaled for production.

S. B. Lawrenson (Lawrenson 2018) has presented a review about the use of green solvents for synthesis in solid phase, in particular, the use of water, greener organic solvents such as THF, acetonitrile, and ethyl acetate, and compressed carbon dioxide in organic and peptide synthesis in solid phase.

An elegant example of Solvent Engineering applied to green processes was presented by Y. Luo and L. Ou (Luo et al. 2023) who proposed a green process for the selective separation and recovery of Li and Co metal ions from spent Li-ion batteries by using a new DES. The DES proposed was formed by betaine hydrochloride and lactic acid. The solvating environment of the metallic ions was modified by using ethanol. The DES ethanol acts as a ligand with strong affinity with the transition metals in solution, leading to a precipitation of Co and allowing the separation from Li that stays in solution. The mechanism through which different transition metals as Co, Li and Ni leach in ethylene-glycol:choline chloride DES from LiB cathode materials has been investigated by S. H. Alhashim et al. (Alhashim et al. 2023). These authors used DFT calculations to elucidate the mechanism and found that ethylene-glycol participates in hydrogen bonding that weakens the metal-oxygen bond, while Cl^- attacks the metal center, Cl^- surrounds the Li, and leaches in the solution. Polyethylene glycol (PEG):phytic acid (PHA) DES were used by Y. Chen et al. (Chen et al. 2023) to leach Co from LiB cathodes. Here it has been observed that the ratio PEG:PHA strongly influences leaching of Co and of Li with the best leaching efficiency at 1:1 ratio. In another work by Q. Yan et al. (Yan et al. 2023), guanidine hydrochloride (GUC) and lactic acid (LAC) based DES modified with 1 wt % of ascorbic acid as reducing agent were used for efficient leaching of $LiCoO_2$ and of lithium nickel cobalt manganese oxides: more than 97% of the Li and 96% of the Co present could have been dissolved. A study by M. Wang et al. (Wang et al. 2023) used a

DES formed by choline chloride and formic acid to extract Li, Ni, Co, and Mn metals from lithium ion batteries cathodes. To enhance the DES efficacy, mechano chemistry was used as well. At the end of the process, lithium was obtained as Li_2CO_3, while Ni, Co, and Mn were recovered as $(NiCoMn)O_x$. The authors claim that this process affords materials that can be used as new precursors for new lithium ion batteries.

The use of mixtures of solvents to control the properties of the solvent medium are unfortunately scarcely reported when greener solvents are intended to be used. Further investigations by using mixtures may represent an excellent opportunity to widen the choice of greener solvents.

4.2 Materials Synthesis and Processing

The possibility to modulate the solvency and transport properties of supercritical fluids, and in particular of supercritical CO_2, has long been used for the fabrication of particles and porous materials. A literature and patent survey concerning years up to 2000 has been published by J. Jung and M. Perrut (Jung and Perrut 2001) about particle design by using supercritical fluids. This review presents the different techniques to manufacture particles, microspheres or microcapsules, liposomes and other dispersed materials such as microfibers by using supercritical fluids. Among these techniques one can find the Rapid Expansion of Supercritical Solutions (RESS), the Gas (or Supercritical fluid) Anti-Solvent (GAS or SAS) technique, the Solution Enhanced Dispersion by Supercritical Fluids (SEDS) and the Particles from Gas-Saturated Solutions (or Suspensions) (PGSS) method. It is interesting to observe across this survey that rationalization of each of the phenomena present in the cited processes is difficult, and that more studies are needed in order to propose predictive tools to choose one or another technique to fabricate materials by using supercritical fluids. Further, these processes have the drawback of solubility of materials in supercritical CO_2, which can be overcome by using more Solvent Engineering in order to develop and design processes capable of yielding materials with the needed properties.

The properties of microporous materials properties can be controlled by controlling the solvent properties. For instance, the work of T. Hasell et al. (Hasell et al. 2014) showed that the solvent, so-called the 'directomer,' used for crystallization, may influence the crystalline polymorph obtained in porous organic cages, allowing control of the connectivity between cage voids and hence the porosity by changing the orientation and packing of the cages through the use of different solvents. The researchers synthesized imine cages by an anti-solvent method, first dissolving the precursor in a good solvent and then adding an anti-solvent to induce crystallization. When the anti-solvent was 1,4-dioxane, the polymorph obtained was different from the one obtained with the other 39 anti-solvents tested. The authors of this study emphasize that the size of the solvent may play a role during crystallization and preferential formation of one of the polymorphs as, if one solvent has the right size to occupy the void space in the cavity formed by two or more cage windows, it will stabilize this cavity and then the entire structure. However, observations performed by testing several anti-solvents of the same size will not result in the same polymorph being formed.

4.2.1 *Synthesis of Perovskites*

The application of Solvent Engineering in the materials synthesis and processing field is well represented in the field of perovskite materials production. Hybrid organic-inorganic metal halide perovskite solar cells have been proposed recently as light absorbers in efficient photovoltaic cells. The advantages of these materials are long carrier diffusion lengths, widely-tunable band gap and great light absorption potential. The low cost of production and high efficiency makes perovskites comparable with Si-based solar cells (Shi and Jayatissa 2018). One of the bottlenecks for the fabrication of perovskite-based solar cells is that simple spin coating from a solution does not yield homogeneous perovskite layers with uniform thickness over a large area. The solutions currently used are prepared in *N,N*–dimethylformamide and *N*-methyl-2-pyrrolidone, among others. Recently, solvent engineering has allowed the fabrication of uniform and dense perovskite layers through the use of a mixture of solvents, γ-butyrolactone (GBL) and dimethylsulfoxide (DMSO), followed by a toluene drop casting, which has enabled the fabrication of remarkably improved solar cells. The mechanism proposed involves an initial stage during spinning, where precursors are dissolved in the mixture DMSO/GBL, then, in an intermediate stage, the concentration of the film is concentrated by the evaporation of GBL, after that, the toluene droplets remove the excess of DMSO leading to a phase where the rest of DMSO retard the rapid reaction between both precursors during the evaporation of the solvent. In the end, the flat film is converted to a crystalline phase after annealing at 100°C (Jeon et al. 2014). In the study presented by J. Zhang et al. (Zhang et al. 2019) Solvent Engineering has been used to obtain two-dimensional (2D) organic-inorganic perovskite materials by using dimethylformamide (DMF) and DMSO together in the precursor solution, allowing the development of high-performant 2D perovskite-based materials, which, when used in the fabrication of solar cells, exhibited a power conversion efficiency of more than 11%. This study shows that Solvent Engineering can help to address issues in 2D perovskite-based solar cells, where one of the bottlenecks is the low performance obtained when a solution of the precursors in only one solvent is used.

Other works in the perovskites-based materials field where Solvent Engineering has been successfully used with encouraging results can be found in references (Ke et al. 2019; Liu et al. 2018; Shi et al. 2019; Tu et al. 2017; Zhang et al. 2018). The importance of Solvent Engineering in this field has been so clear that the dynamics occurring during the solution processes involved have been discussed in a publication by Z. Arain et al. (Arain et al. 2019), where the capacity to dissolve the precursors, boiling points and Lewis acid-base character of the solvent are discussed in each case. The authors propose a framework for a judicious choice of a solvent system for obtaining the desired perovskite solution. Other green solvents that have been used for the fabrication of perovskite precursors or to deposit perovskite films include DMSO (Noel et al. 2022), water and anisole together with DMSO and dibuty ether (Cao et al. 2023; Wu et al. 2023), γ-valerolactone and *n*-butyl acetate (Miao et al. 2023), (R)-(+)-limonene and 2-methyltetrahydrofuran (Hyun Park et al. 2021), 2-methylanisole (2-MEA) (Ma et al. 2023; Yu et al. 2023), ethyl acetate (Onozawa-Komatsuzaki et al. 2023) and ionic liquids (Wang et al. 2023) among others.

A solvent-polishing method has been proposed by H. Wang et al. (Wang et al. 2023) by using the green solvent, 2,2,2-trifluoroethanol (TFEA), to polish and reconstruct the quasi-2D perovskite film surface, after which the surface was revealed with less defects and better smoothness. Salicylaldehyde has been proposed as non-toxic and non-halogenated solvent coming from buckwheat as a multi-functional solvent: non-polar anti-solvent and polar post treatment-dissolving solvent. This compound has a benzene ring and a hydroxyl group allowing it to be used as non-polar and polar compound for perovskites solar cells fabrication (Cho et al. 2023). An interesting strategy to decrease up to 70% of the toxic waste and material cost by using Solvent Engineering is the one proposed by H. Zhang et al. where the use of a judicious co-solvent to alter the colloidal physical chemistry of the precursor allows a decrease in the concentration by obtaining the same high-quality perovskite films as when high concentrations are used (Zhang et al. 2022).

S. K. Podapangi et al. have prepared a review gathering newly-published scientific articles about green solvents for sustainable fabrication of perovskite solar cells (Podapangi et al. 2023). A comprehensive and pertinent discussion about the use of different solvent scales and their advantages and limitations for replacement of toxic solvents by greener solvents for perovskites applications in solar devices can be found in this review. Several other review articles have been published concerning solvent use, impact and engineering in perovskites synthesis and processing as those found in references (Kwon et al. 2023; Cheng 2023; Jiao et al. 2023; Xiaobing Cao et al. 2019; Arain et al. 2019).

Another subject in the materials field where solvent effects have attracted great attention is the fabrication of hydroxyapatite, as this material finds potential industrial applications in medicine, biology and fertilizer production. Prepared through a solvo-thermal process, the physico-chemical properties of the solvent, such as the dielectric constant, the interionic attraction and the solute-solvent interactions strongly influence the solubility and diffusion behavior of the chemical precursors in the solvent. For instance, Jung et al. (Jung et al. 2009) used water and a water/solution to perform the synthesis of hydroxyapatites through a spray pyrolysis method and studied the effect of the solvent on the physical properties of the materials obtained. They observed that larger size and bimodal particles were formed when the substrates were dissolved in water, while monodisperse and nano-sized particles were produced when a mixture of water/ethanol was used.

M. Anwaar et al. (Anwaar et al. 2019) studied the effect of polarity and of dielectric constant of the solvent on the calcium phosphate phases obtained in the materials synthesized by a sol-gel process. They used pure water, water/DMF and water/tetrahydrofuran (THF) as solvents. When water alone, with the highest dielectric constant, was used, large β-calcium pyrophosphate (β-CPF) plate-like structures with a minor amount of hydroxyapatite were obtained. DMF/water yielded hydroxyapatite as the major phase with minor amounts of β-CPF, and the use of THF/water, which was the solvent system with the lowest dielectric constant, resulted in the formation of a pure hydroxyapatite phase.

4.2.2 Electrolytes

Electrolytes are ubiquitous in industry in carbon capture, storage and utilization, batteries technology, biotechnology, and so on. Liquid electrolytes are often obtained by solubilizing a salt in a solvent and solvent effects can be dramatic in the properties and in the performance of the obtained electrolyte (Pabsch et al. 2022). Organic solvents and co-solvents have been tested to improve the performance of liquid electrolytes as, for instance, in Lithium-ion batteries including esters and ethers. One of the challenges in liquid electrolytes is to decrease the viscosity to enhance the mass transport and hence the Li ion conductivity of the electrolyte (Wu et al. 2020). Besides transport of the ions, the interfacial chemistries on the electrode surfaces are greatly influenced by the solvent used in the electrolyte, and synergistic effects between the salt and the solvent when they coexist on electrode surfaces may be triggered if the electrolyte is judiciously formulated (Xia, Cao et al. 2021). Electrolytes are then a field where Solvent Engineering is useful and has been performed for some time.

In particular, in recent years, localized high-concentration electrolytes (LHCEs) have emerged as promising alternatives to previous electrolytes because of their stability and their interesting solid electrolyte interphase

(SEI) formation mechanism. A typical LHCE is formed by a conducting salt, a solvating (or ionizing) solvent, and a non-solvating diluent (or non-ionizing solvent). Solvents typically used as ionizing solvents are 1,2-dimethoxyethane (DME), dimethyl carbonate (DMC), trimethyl phosphate (TMPa), tetramethylene sulfone (TMS), and acetonitrile (AN). Electrolyte additives are currently ethylene carbonate (EC), vinylene carbonate (VC) and fluoroethylene carbonate (FEC). A currently used diluent is 1,1,2,2-tetrafluoroethyl-2,2,3,3-tetrafluoropropyl ether (TTE). H. Jia et al. have systematically investigated the effects of solvating solvents and additives in the performance of LHCEs (Jia et al. 2023). Other studies concerning solvent effects on LHCEs performance can be found in literature (Peng et al. 2022; Ding et al. 2017). Several studies have proposed Gutmann's Donor Number (DN) as one of the key parameters to predict the effect of the solvent on the electrolyte. It is for instance well known that high DN solvents present catastrophic reactivity with a Li metal anode, such as DMSO, DMA, DMF, etc., which limits their practical application in Li–S batteries. Some authors have then proposed the use of high-DN solvents as additives instead of as co-solvents for conventional ether electrolytes (Zhong et al. 2022; Gupta et al. 2021). A judicious balancing of the DN has been a strategy proposed by P. Zhou et al. (Zhou et al. 2023), who used an electrolyte formed by $LiNO_3$ and DMI, a urea-based molecule with moderate DN presenting high solubility for $LiNO_3$; this electrolyte allowed a Coulombic efficiency close to 100%. A. W. Tomich et al. (Tomich et al. 2024) have proposed an innovative Mg-based electrolyte in DME by engineering the surface of the Mg compound.

In an effort to propose a model for electrolytes in systems using solvents other than water, Cameretti et al. (Cameretti et al. 2005) proposed the Equation of State ePC-SAFT (electrolyte perturbed-chain statistical associating fluid theory), which was later revised by Held et al. (Held et al. 2014). ePC-SAFT has been successfully applied to battery electrolytes and can be used to predict the properties of an electrolyte where the solvent has been judiciously engineered to meet the needs of next generation devices. An interesting variant of SAFT for electrolytes was presented by N. Novak et al. and is called the eSAFT-VR Mie. A review of the most used models for electrolytes is also presented in this publication (Novak et al. 2023).

4.2.3 Polymeric Membranes

One domain where the solvent effect is omnipresent and consequently studied is the fabrication of polymeric membranes. The fabrication of polymeric membranes is widely performed by solvent exchange or, in other words, by diffusion induced phase separation, where a mixture of a polymer and a solvent is cast as a thin film and immersed into a coagulation or quenching bath containing a non-solvent of the polymer. The non-solvent contained in the coagulation bath needs to be miscible with the solvent of the initial solution. The entire process is governed by the diffusion of the various components, which naturally compels the performance of Solvent Engineering to optimize the process. The effect of the solvent and the non-solvent on the obtained materials has been the object of several research studies like the one presented by Barton et al. (Barton et al. 1997) where polyethersulfone (PES) membranes were prepared in two ternary systems: PES/DMSO/water and PES/NMP/water. An *in situ* dynamic measurement system was set up by the authors to observe the dynamics of the membrane formation process in the polymer/solvent casting film. These observations showed that rapid precipitations result in finger-like macrovoids formation, while delayed precipitations yielded spongy morphology. This dynamics can be controlled by the quench bath composition. In this study, the quenching bath composition was composed of H_2O/DMSO v/v of 100/0 and 10/90. In the first case, the quenching was rapid with finger-like macrovoids, while in the second it was delayed with a spongy structure. Interestingly, Smolders et al. (Smolders et al. 1992) have suggested that macrovoid growth is controlled by diffusion flows, while McKelvey and Koros (McKelvey and Koros 1996) have proposed that macrovoid formation is driven by osmotic pressure forces generated in the nucleated polymer-lean phase.

Several mathematical models of the phase inversion process have been published (Cohen et al. 1979; Hopp-Hirschler and Nieken 2018; Radovanovic et al. 1992a, 1992b; Reuvers and Smolders 1987; Strathmann et al. 1975; Tsay and Mchugh 1990; Wijmans et al. 1985; van de Witte et al. 1996). The models presented by Reuvers et al. (Reuvers et al. 1987) and by Smolders et al. (Smolders et al. 1992) discussed the thermodynamic aspects of instantaneous and delayed demixing processes. In the first case, the polymer precipitates quickly in the coagulation bath, which generally yields finger-like macrovoids with thin skin layers; in the case of delayed demixing, precipitation takes a long time and the membrane presents a sponge-like structure with a dense top layer. Fast solvent-non solvent exchange occurs when high miscibility exists between the solvent and the non-solvent. On the contrary, low miscibility between the solvent and the non-solvent yields slow solvent- non solvent exchange and so the exchange is delayed. The work by Tsai et al. (Tsai et al. 2010) discussed the effect of solvent quality on membrane morphology. They used two solvents: N-methyl pyrrolidinone (NMP) and 2-pyrrolidinone (2P) to fabricate membranes in polysulfone (PSf) using water in vapor form for a certain time and then in liquid form as a non-solvent. N-methyl pyrrolidinone yielded structures with macrovoids, produced by the fast demixing of the ternary system, while 2-pyrrolidone, which is a poorer solvent for

(a)   (b)

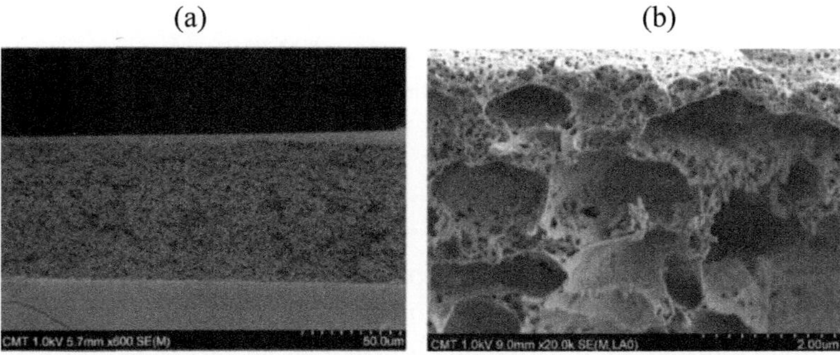

Figure 16. Dependence of the PSf cross-sectional structure on the solvent used (a) 2-pyrrolidinone (12%wt PSf), and (b) NMP (20%wt PSf) after 3 minutes of exposure to vapor and then to liquid water as coagulation bath.

PSf, afforded spongy structures, obtained by the slow demixing of phases (Figure 16).

The work of Yeow et al. (Yeow et al. 2004) about the fabrication of poly(vinylidene fluoride) membranes (PVDF) in four different solvents: N,N, Dimethylacetamide, N,N, dimethylformamide, 1-methyl-2-pyrrolidon, and trietyl phosphate, and, using water as coagulation bath, corroborated that a weak solvent yields a spongy structure, while macrovoids are obtained when a stronger solvent is used.

The use of additives in the casting solution is ubiquitous when polymeric membranes are prepared. These additives can be separated into

1) polymeric additives, such as polyethylene glycol of different molecular weights,

2) weak non solvents, such as glycerol,

3) weak co-solvents such as ethanol and acetone, and

4) low molecular weight inorganic salts such as lithium chloride and lithium perchlorate (Yeow et al. 2004).

As can be easily imagined, the effect of these additives on the resulting membrane varies in different polymer/solvent/non solvent systems. These additives need to be taken into consideration when a modeling of the system is performed in order to predict the structure obtained at the end. The effect of additives in different polymeric membrane structures has been studied by several research groups for polymers such as polyvinyl alcohol (Chuang et al. 2000), polysulfone (Kim and Lee 1998; Urkiaga et al. 2015), PVDF (Kaner et al. 2019; Shen et al. 2019), and a poly(dimethylsiloxane) (PDMS) – poly (methyl methacrylate) (PMMA) (Miyata et al. 2001; Uragmi et al. 2006) among others.

The preparation of polymeric membranes has for long used Solvent Engineering activities, as the fabrication of different structures directly depends on the properties of the solvents and of the non-solvents used. Researchers and industry have tried to optimize the systems in order to obtain spongy, mechanical resistant and homogeneous structures, which have necessarily gone through an optimization of the solvent properties. It is to be noted that recently, green, biodegradable solvents have attracted great attention as substitutes for toxic non-biodegradable solvents for the fabrication of polymeric membranes. In the work performed by Y. Medina-Gonzalez et al. (Medina-Gonzalez et al. 2011), cellulose acetate membranes were fabricated by using alkyl lactates as solvents and water as the coagulation bath. The review published by A. Figoli concerning non-toxic solvents for membrane preparation is highly recommended to interested readers (Figoli et al. 2014). In particular, efforts to avoid the use of toxic, environmentally-harmful solvents such as N-methyl pyrrolidinone (NMP) to obtain membranes of similar quality are strongly needed.

N. Naderi et al. (Naderi et al. 2022) have described the effect of solvent mixtures on the morphology and performance of polyacrylonitrile ultrafiltration membranes. For that, dimethylformamide (DMF) and N-methyl-2-pyrrolidone (NMP) mixtures were used as solvents for the polymer and, once the polymer solution was cast, a coagulation bath filled with water was used to perform a solvent exchange and to form the membrane by the technique called non-solvent-induced phase separation (NIPS) which is one of the most used strategies to produce asymmetric membranes. During NIPS, macrovoids are often produced and their suppression is one of the greatest challenges in membranes production as they decrease the structural and mechanical stability of the membrane. These macrovoids are often suppressed by adjusting the formulation of the dope solution, which is formed by solvents or mixtures of solvents, polymer and other compounds. The coagulation bath can also be formed with non-solvents and can include solvents of the polymer. The formulation of the coagulation bath plays a role in the formation of macrovoids and of pores at the surface of the membrane. In this study, by changing the composition of the polymer solution, different microstructures were obtained, going from macrovoids to sponge-like structures. In conclusion, the authors stated that both the polymer concentration and the solvent ratio played an important role in the improvement of the membranes structure and properties and of their performances.

4.3 Soft Matter

Solvent effects observed in soft matter fields are very varied, and can range from inducing rheological changes in gels to influencing the collapse of polymer chains or promoting specific self-assembled arrangements into supra molecular structures. Solvent effects have been related to macroscopic (refractive index, density, etc.) and microscopic properties (solvation, intermolecular forces, etc.). For instance, in assembling reactions, the solvent medium determines activation energies and reaction thermodynamics, changing potential energy barriers and, as a consequence, the free energy, by modifying the solute-solvent interactions. Some examples of solvent effects on soft matter will be presented in the following paragraphs, co-non solvency phenomena and solvent effects on active systems will be developed as well.

C-P. Li and M. Du (Li and Du 2011) have prepared a review about the solvent effects in coordination supra molecular systems, taking into consideration the role of solvent as ligand, as guest, where steric effects, H-bonding capabilities, hydrolysis effects and solvent-induced polymorphism may be present. A part of this work is devoted to the role of solvents in the obtained properties of the coordination supra molecular systems, such as the effect on gas storage and separation, on catalysis, on magnetic and on optical properties.

One of the most interesting examples of the solvent effect and more specifically the effect of the addition of a co-solvent, is the co-non solvency effect, which is an abnormal response of polymers that normally swell in good solvents and collapse in a poor solvent, but that in a mixture of good solvents collapse. This is a phenomenon that has intrigued polymer and physical-chemistry scientists who have proposed it as a stimuli-responsive effect that may have practical applications in, for instance, drug delivery systems and in sensor devices. Different research groups have worked on this phenomenon and on its rationalization with the aim of predicting which systems may present it. Some of the systems widely studied are poly(N-isopropylacrylamide) (PNIPAM) in alcohol/water mixtures (Delbecq et al. 2020), PNIPAM in other mixtures of solvents such as THF/water (Mukherji et al. 2016; Saunders et al. 1997; Walter et al. 2012; Winnik et al. 1992; Yu et al. 2016) and even other polymers and copolymers in organic solvent/water mixtures (Hao et al. 2010). It has been shown in different works that a good choice of solvents can tune the polymer conformation (Colja et al. 2018; Hossein Mahdavi et al. 2009; Orakdogen and Okay 2006; Pagonis and Bokias 2004). An interesting publication by D. Mukherji et al. (Mukherji et al. 2016, 2019) presents a review about co-non-solvency and co-solvency phenomena, the latter being associated to the swelling of a polymer or co-polymer in poor solvent mixtures. An example of this effect is poly (methyl methacrylate) (PMMA) in aqueous alcohol mixtures. While pure alcohol and pure water are poor solvents for PMMA, a mixture of water and alcohol improves the solubility of this polymer. Other polymers have shown this behavior too, such as corn starch or poly(N-(6-acetomidopyridin-2-yl)acrylamide); PMMA also shows this behavior in mixtures of 2-butanol and 1-chlorobutane. Co-non-solvency and co-solvency were discussed by the authors within unified generic concepts common to both phenomena.

Theoretical studies on the effect of co-solvents on polymer swelling and collapse have been published by some research groups. For instance, A. Muzdalo et al. (Mukherji et al. 2017) have presented a study of the collapse and swelling of a generic homo-polymer by an approach that used implicit-solvent, explicit-co-solvent Langevin dynamics computer simulations. These calculations have revealed that polymer swelling in the presence of a co-solute is maximal if both monomer-monomer and monomer-co-solute interactions are weakly attractive. Highly attractive monomer-co-solute interactions induce a collapse of the chain, which can be observed in purely repulsive co-solvents as well. Different mechanisms can thus induce collapsed structures, with different thermodynamic and structural properties. Various mechanisms have been proposed to explain the phenomenon of co-non-solvency, for instance the cooperativity effect (Heyda et al. 2013), solvent–co-solvent interactions (Kojima et al. 2013; Tanaka et al. 2011; Tanaka et al. 2008), preferential co-solvent–monomer interactions (Dudowicz et al. 2015; Jia et al. 2016) and the kosmotropic effect (Backes et al. 2017; Heyda et al. 2013; Magda et al. 1988; Mukherji et al. 2014; Winnik et al. 1992).

Mean-field approaches have been shown to correctly describe co-non-solvency in some cases, as shown by different authors combined with the preferential adsorption concept as in PNIPAM microgels in DMF/water solutions (Bischofberger et al. 2014). Other theoretical studies have suggested preferential absorption of alcohol by strong interactions with the polymer chains when it is in contact with alcohol-water solutions, while the behavior is the opposite in DMSO-water solutions, where the co-solvent is excluded from the collapsed, polymer-rich phase (Zhu and Chen 2019). In the case of alcohol-water solutions, evidence has been found that alcohol species are accumulated inside the polymer network in its collapsed region, and this behavior is more pronounced with increasing hydrophobicity of the alcohol moiety (Nothdurft et al. 2019). In the study by T. Zuo et al., MD simulations have shown that co-non-solvency is a result of strong water-co-solvent attraction (Nothdurft et al. 2019).

Active matter systems, such as colonies of bacteria and self-propelled synthetic micro swimmers are a vast field of study of soft matter. These are self-driven systems that live or function far from equilibrium, because their microscopic constituents constantly consume energy from the surrounding environment in order to do work, for instance to propel themselves. Intriguing phenomena arise from the non-equilibrium nature of active systems (Zuo et al. 2019). These characteristics of self-propulsion allow these systems to be free of classical thermodynamic constraints, enabling them to control their motion and the environment in which they evolve (Gonnella et al. 2015). Non-equilibrium suspensions of active particles that constantly convert energy into directed motion are promising in self-assembly (Takatori and Brady 2015) and targeted cargo transport (Prymidis et al. 2015; Wensink et al. 2014) among other applications. Solvent interactions and effects on active matter have begun to be studied, as in the theoretical work by J. Rodenburg et al. (Baraban et al. 2012; Snezhko and Aranson 2011) where the authors have generalized Van't Hoff's law for active suspensions. It has been evidenced that active particles exert a net reaction force on the solvent that can be experimentally measurable as a solvent flow through a semi-permeable membrane confining the active suspension to one side of the open pipe or a macroscopic rise of the suspension meniscus in a *U*-pipe experiment. In the latter case, this reaction force increases the solvent pressure of the suspension and hence the solvent chemical potential (Rodenburg et al. 2017).

A phenomenon that is unique to active systems is 'motility-induced phase separation' (MIPS). This term refers to the phase separation into a concentrated and gas-like diluted phase observed in a system of self-propelled particles that interact only via steric or excluded volume repulsion, where the overall density is large enough (Rodenburg et al. 2017). In other words, MIPS can be understood as follows: if the local density of self-propelled particles increases in some part of the system, particles in this region will slow down due to enhanced crowding. Self-motile particles accumulate where they move more slowly. This behavior can trigger a feedback, where the particles accumulate where they are slower, further slowing down the movement due to crowding, and then accumulate even more, etc. (Gonnella et al. 2015). The effects of solvent on MIPS are still unclear, as solvent-mediated hydrodynamic interactions in the context of MIPS have not yet been thoroughly studied. The presence of these interactions affects the dynamic properties of soft matter in general and in particular of active systems: they can change the diffusion and rheological properties of the suspension and hence affect the motility of active particles. Some theoretical works have evidenced the effect of these interactions on motility of active matter. However, these studies have focused on the physical properties of solvents, while the chemical properties of solvents and their effects on motility of active particles have yet to be discussed (Gonnella et al. 2015).

4.4 Reaction Chemistry

Solvents significantly influence the kinetics and thermodynamics of chemical reactions, as well as product selectivity. The influence of solvents can be observed in

1) direct participation in the reaction steps and in reaction pathways,
2) competition with reactants for interaction with the catalyst,
3) modification of the stability of reactants, transition states or products,
4) alteration of diffusion properties, notably of reactants and products through pores of the catalyst,
5) change in free energy barriers through entropic confinement,
6) changes in solubility of the different compounds and
7) inhibition of undesired reactions.

Generalization of solvent effects and, even more, an extrapolation of these generalizations to design solvents to optimize chemical reactions or to promote a desired pathway or product, is a challenging task that has interested many research groups for a long time. A recent review showing some solvent effects on different chemical reactions such as addition of radicals to multiple bonds, β-scissions and Diels-Alder reactions, among others, has been published by B. L. Slakman and R. H. West (Bär et al. 2020; Marchetti et al. 2013), with an emphasis on computational methods used to study the solvent effects on the cited reactions. Finally, some references about the recent use of machine learning for parameter prediction based on molecular properties are given.

A comprehensive discussion on solvent effects in catalysis has been presented in the document by J. J. Varghese and S. H. Mushrif (Slakman and West 2019); axed on heterogeneous catalysis, which implies phenomena at the liquid/solid interface. It has a section devoted to computational tools and methods available to investigate these effects. There are sections concerning the origins of changes in the activity and/or durability of the catalyst induced by the solvent; the origins of the alteration of product selectivity by solvent effects; and even the inhibition of undesired reactions by reactive solvation, and interested readers are encouraged to consult these. Computational tools are discussed from two points of view: implicit solvation and explicit treatment of solvents, with a concise and comprehensive treatment of the methods used on both approaches.

Among the indispensable oeuvres discussing solvent effects on chemical reactions are, without the aim of being exhaustive, the book proposed by C. Reichardt about the effect of solvents on the rates of homogeneous chemical reactions and the book (Varghese and Mushrif 2019) about solvent effects in organic chemistry.

Chemical reactions represent a challenge when the aim is to rationalize the effect of solvents. Solvent Engineering for a specific chemical reaction has been done by using co-solvents and mixtures of solvents. However, rationalizing co-solvent effects is more difficult and can be even more complicated if the chemical reactions studied are complex. Transesterification reaction has attracted attention to produce biodiesel from triglycerides and an alcohol such as methanol, with a strong base as catalyst. This reaction presents a miscibility problem as methanol and triglycerides are not miscible. To overcome this problem, this reaction can be performed at high temperature or by using microwaves, both of which are highly energy-consuming. Another strategy is to use a co-solvent to increase miscibility. Some co-solvents that have been used are n-hexane, diethyl ether, acetone, 2-propanol, tetrahydrofuran, and ethyl acetate. In this context, T. S. Julianto and R. Nurlestari (Christian Reichardt and Welton 2010) presented a study of the effects of the concentration of acetone as a co-solvent in transesterification reaction. Acetone has intermediate polarity which makes it miscible with methanol and triglycerides, allowing a homogeneous reaction media improving the rate of transesterification. Acetone also stabilizes the active intermediate compound, the methoxide ion which then attacks triglycerides to form methyl esters. Acetone has increased transesterification yield when used at a ratio acetone:methanol of 1:2, this ratio has been optimized, as ratios of 1:4 or 1:1 decrease the yield.

In a work developed by B. Sanchez et al. (Julianto and Suratmi 1823) solvent effects were studied on a nucleophilic aromatic substitution reaction (S_NAr), for instance the S_NAr reaction between 2,4,6-trinitrophenyl ether (PTNPE) and piperidine in a solvent formed by mixtures [BMIM][BF_4]-water at different concentrations of water. From this study, it can be concluded that controlling the water content in the solvent mixture allowed control of the kinetics of the reaction, where low concentrations of the ionic liquid promoted the reactivity of the substrate through preferential solvation phenomenon by hydrogen bonding, while at high concentrations of the ionic liquid, a catalytic effect from the ionic liquid was observed.

P. R. Campodónico (Sánchez et al. 2018) has published a review about solvent effects in S_NAr reactions in conventional and non-conventional solvents. Here, one section is devoted to the effect of co-solvents on this type of reactions either in mixtures of water with ionic liquids or in mixtures of other solvents. Juan A. Araque et al. (Araque et al. 2015) have studied by molecular dynamics the micro viscosities of different media, in particular mixtures of DMSO and glycerol with different concentrations of glycerol and ionic liquids solvents. During simulations of DMSO/glycerol, nanoscopic segregation between DMSO-rich domains and hydrogen bonded glycerol networks was found.

T. W. Walker et al. (Maciejewska 2020) have examined the effect of solvents and co-solvents on acid-catalyzed reactions of biomass-derived oxygenated molecules such as ethyl *tert*-butyl ether, *tert*-butanol, levoglucosan 1,2-propanediol, fructose, cellobiose and xylitol in solvent mixtures of water with three polar aprotic co-solvents: γ-valerolactone, 1,4-dioxane and tetrahydrofurane. The authors used classical molecular dynamics simulations to observe the extent of water enrichment in the local solvent domain of the reactant, the hydrogen bonding lifetime between water molecules and the reactant and surface of the reactant occupied by hydroxyl groups. The authors emphasize in their study the purpose of deliberately driving the formation of water-enriched local domains near hydrophilic reactants by tuning reactant-solvent-co-solvent interactions. A higher enrichment of water near the reactant yields reaction rates despite differences in reactant hydrophilicity and reaction mechanisms. However, the other two parameters needed to be taken into account to fully predict the outcome of the use of co-solvents in these reactions, so it is a combination of the three proposed parameters in a multi-parameter correlation that predicts rate constant as a function of solvent composition. This interesting strategy allows control of reaction rates by controlling solvent properties by adjusting the concentration of the co-solvent.

Using MD calculations and experimental determinations, B. Mostofian et al. (Walker et al. 2018) studied the effect of a solvent, in particular of mixtures THF-water, in the hydrolysis of biomass components to sugar monomers. They observed the effect of preferential solvation of glycosidic bonds by water when THF is present, and noted the symbioses of the THF-water system, where the differential solvation between both solvents promotes solubilization of cellulose and reaction in sulfuric acid solutions in THF-water as compared with aqueous solutions of sulfuric acid.

Concerning lignocellulosic materials dissolution, P. Dominguez de Maria (Domínguez de María 2014) prepared a mini-review about the use of neoteric solvents for the dissolution of these materials. One of the main hindrances in cellulose solubilization is the crystallinity and the recalcitrance of cellulose. The review includes CO_2 switchable ionic liquids such as DBU (1,8-diazabicyclo[5.4.0]undec-7-ene), distillable ionic liquids that combine dimethyl amine and CO_2 (2:1 eq/eq) that triggers the formation of dimethylammonium ion and dimethylcarbamate ion. These ionic liquids can be distilled at 45°C, which is suitable for their recovery. Choline-based ionic liquids and deep eutectic solvents (DES) are addressed as well; in particular choline-based ionic liquids that are considered as bio-based have been thoroughly used for cellulose dissolution as in the case of cork biopolymers dissolution and hemicellulose (xylan) and lignin dissolution. In the case of DES, an *in situ* and natural formation of DES formed by malic acid and sucrose has been observed during biochemical reactions. Judicious engineering of the DES has allowed obtaining a dissolution of lignin in quaternary ammonium-based and in amino acid-based DES.

Lignin dissolution in environmentally friendly, non-toxic, and cost-effective solvents is a great challenge that should be addressed in a bio refinery-based economy, as the production of lignin is estimated to be 100 million tonnes/year (Bajwa et al. 2019) and it can be used to produce phenolic-containing molecules and macromolecules and aromatic compounds (da Cruz et al. 2022; Shuai et al. 2016; Subbotina et al. 2021; Sun et al. 2018; Weng et al. 2021; Chaofeng Zhang and Wang 2020). M. G. A. da Cruz et al. (da Cruz et al. 2022) have proposed the use of levulinic acid to solubilize/depolymerize lignin to form aromatic monomers and dimers by using an electrochemical route and copper as the electro-catalyst.

One of the best examples of the extent of Solvent Engineering possibilities in reaction chemistry may be the approach presented recently by C. Zhang et al. (Mostofian et al. 2016), with the objective to perform multistep reaction-purification processes by smart solvents, and without the need for intermediate separation and purification modules. Here the solvents used need to support and/or perform the functions of reactions and separation in an integrated manner, called 'solvent factory' in order to use solvents as an integrated reactor-separator system. The solvent selection method proposed by these authors is based on the Conductor like Screening Model for Real Solvents (COSMO-RS). The choice of the solvent would aim at enabling each solvent to be the exclusive or prime host of a guest species, for instance, one solvent for the reactants (the best one for each reactant), another one for the catalyst and one for the product. In this manner, the solvents have the function to host, release and finally trap the species for recycling and products purification. This approach has been exemplified with the aldol reaction. For that, after choice of the selected solvents, a triple phases system was obtained, formed by dodecane, water and reaction products, where product separation was easy. A micro-flow reactor was designed in this study to improve mixing and to increase the contact surface for the catalyst.

4.5 Biochemistry

Solvation effects are the driving force in various macro-molecular processes, from those found in hydrogels responsive to external stimuli to smart polymers responsive to solvent concentration gradients or even in denaturation of proteins (Zhang et al. 2020). Some biopolymers show the phenomenon of co-non-solvency, which refers to the collapse of a polymer in a mixture of two good solvents, as in the case of elastin-like polypeptides that are water-soluble, but present a coil-to-globule-to-coil transition in an aqueous solution of ethanol (Zhao et al. 2020). This behavior is alcohol concentration dependent, which suggests that a solvent can allow a tunable phase behavior of these biocompatible polymers, giving access to advanced stimuli-responsive materials.

Biochemical reactions can be modulated by solvent additives such as co-solvents, if, for instance, during the course of the reaction, there is a change in preferential interactions of solvent components with the reacting system that can be either preferential binding or preferential exclusion. S. Timasheff (Zhao et al. 2020) has presented the basic thermodynamics of binding that define the nature of the processes existent between ligands and water molecules interacting with a biological system.

Solubility or aggregation of a protein in a solvent, that can be water or an organic solvent, is at the heart of many biochemical, biotechnological and medical applications such as the development of protein formulations used in medicine or the appearance of diseases such as Parkinson, Alzheimer or even cataract formation. It therefore follows that efforts devoted to the study of the solubility of proteins and means of predicting and controlling the phenomena involved in this process were studied during recent decades by several research groups (Timasheff 2002; Middaugh et al. 1979; Wang and Ben-Naim 1997). Together with the solvent, co-solvent effects have been studied as a stimuli capable to modulate the stability of protein conformations by influencing the kinetics of aggregation. Aggregation, either in globular or in intrinsically disordered proteins, can lead to mature amyloid fibrils, which have been linked to neuro-degenerative disorders, and have been observed to be affected by, among other physico-chemical parameters, by co-solvents such as osmolytes. B. Mondal and G. Reddy have studied the effect of six co-solvents on fibril growth of a model protein through a coarse-grained protein model and computer simulations (Tournier et al. 2003). Their conclusions are that denaturant co-solvents decreased the energy barriers for dimers formation, while protective co-solvents increased these barriers. Heat stability of proteins has been observed to increase in solutions containing sugars such as sucrose, glucose (Mondal and Reddy 2019) or even in the presence of polyols such as glycerol or sorbitol (Lakshmi and Nandi 1976). Competing solvent effects between the polyols and the denaturant were observed, and the polyols seemed to stabilize the protein structure through strengthening of the hydrophobic interaction.

The effect of solvation on the conformational preferences (e.g., R-helix versus â-sheet) of tripeptides has been studied by C. Park et al. (Gekko and Ito 1990) by using *ab initio* quantum mechanics with solvation in the Poisson Boltzmann continuum solvent approximation. These authors have found that α-helix conformations are preferentially stabilized over β-sheet conformations by aqueous solvents.

Several reviews published about solubilization and stabilization of proteins in solution, and solvent and co-solvent effects have attracted attention during, at least, the last 30 years. Without the intention of being exhaustive, some of the reviews can be cited: the review by C. H. Schein (Park et al. 2000) presents a good synthesis about ways to stabilize proteins as well as some methods to determine, predict, and increase their solubility. Solvent additives (osmolytes) that stabilize proteins are treated with a description of their effects on proteins and on the solvation properties of water. More recently, O. Tapia (Schein 1990) proposed a review about solvent effect theories, in particular quantum and classical formalisms and their applications in chemistry and biochemistry. M. Orozco and F. Javier Luque (Tapia 1992) published a review about the theoretical methods used for the study of solvent effects in biomolecular systems. D. Roccatano (Orozco and Luque 2000) prepared a review about non-aqueous solvent effects on biomolecules studied by computer simulations. Concerning reactions, solvent effects play a determinant role. A. Kafle and L. Cui have presented a review about solvent effects on the stereo-selectivity of glycosylation reactions (Roccatano 2008), which is a special topic in carbohydrate chemistry.

The effect of solvents on the activity of enzymes is, as well, a vast topic addressed in recent decades. Some reviews on this topic can be cited, such as those prepared by M. N. Gupta (Kafle et al. 2016), by H. Ogino and H. Ishikawa (Gupta 1992), by S. Wang et al. (Ogino and Ishikawa 2001), and by A. Kumar et al. (Wang et al. 2016).

It is to be noted that enzymes have shown astonishing behaviors in solvents, such as catalyzing reactions impossible to be developed in water, improved stability, and control of the selectivity or even inversion of it in organic solvents (Klibanov 2001; Wescott and Klibanov 1994). It has, however, been found as well that in some cases, organic solvents decrease the activity of enzymes. In such cases, Solvent Engineering has been used to improve their performance as the review by Y. L. Khemelnitsky et al. (Kumar et al. 2016) shows. Here the compounds used to improve catalytic properties of the enzymes are carbohydrates, polymers and organic buffers. This reveals the importance of the solvent micro environment on the activity of the enzyme and emphasizes that a judicious control of this micro environment can conduct improved or even unattended activities and even selectivity. The review by V. Stepankova et al. (Khmelnitsky et al. 1994) compiles the strategies of stabilization of enzymes in organic solvents.. One of the strategies presented is the modification of the solvent environment to decrease its denaturing effect on enzymes. Surprisingly ionic liquids and deep eutectic solvents can be used for this strategy. More specifically, T. Gerhards et al. (Stepankova et al. 2013) have published a systematic study about the influence of organic solvents on enzymatic asymmetric carboligations of aldehydes with thiamine diphosphate

(ThDP)-dependent enzymes. By using the carboligation of acetaldehyde and benzaldehyde as reaction, six ThDP-dependent enzymes and 13 co-solvents in different concentrations; the authors emphasize the potential of medium engineering, what we can call Solvent Engineering, as a powerful additional tool for varying enzyme selectivity and thus engineering the product range of bio transformations. They however stress that the use of co-solvents should be carefully planned, as the solvents may compete with the substrates for binding sites in the enzyme active site.

Biocompatibility of ionic liquids has been a subject of recent interest. In particular, choline-based ionic liquids have been studied for their biocompatibility for the storage and stability of proteins and biomolecules and as drug delivery agents. Choline-based ionic liquids present a strong affinity with lipid bilayers, cytochrome-c, lysozyme and nucleic acids, which contributes to the stability of the ensemble. In this context, K. D. Tulsiyan et al. (Devi Tulsiyan et al. 2023) observed the stability and structure of hemoglobin in choline-based ionic liquids. The ionic liquids studied led to the partial unfolding of the protein, and present strong hydrogen-bonding interactions.

E. Catoni et al. (Catoni et al. 1996) used the term «Solvent Engineering» in a work about lipase-catalyzed esterifications, where a systematic study of this reaction in many different solvents with different physico-chemical characteristics was performed by using a microbial lipase from *Pseudomonas* sp., immobilized on silica-gel as catalyst. A tendency can be observed from this study where higher dielectric constants and a higher Hildebrand solubility parameter of the solvent decrease conversions, while a higher log P_{ow} will favor the conversion that can be explained by the ability of the solvent to distort the water layer present around the enzyme.

4.6 Supra-molecular Systems

Solvents and solvation play a fundamental role in supra-molecular systems, where non-covalent interactions such as van der Waals interactions, hydrogen and halogen bonding, electrostatic interactions, π-π stacking, charge-transfer interactions, metal-ligand coordination, ionic interactions and hydrophobic or solvophobic effects contribute to form molecular-engineered compounds from small molecular building blocks. The strength of these interactions is in typically the range of 10 to 100 kJ/mole, instead of the hundreds of kJ/mole found in covalent bonds. The dynamic behavior of these interactions may afford stimuli-responsive, self-healing and easy recyclable supra-molecular structures. In marked contrast to covalently formed materials, supra-molecular systems issued from non-covalent interactions present solute-solvent interactions of similar strength than solute-solute interactions, and in consequence, the structure and properties are even more determined by the solvent environment than in covalent systems (Gerhards et al. 2012).

The solvation environment is able to modify the thermodynamics and the kinetics of these complex systems and these effects have been evidenced in systems such as supra-molecular architectures, supra-molecular polymers and complexes and in molecular machines, as for instance in rotaxanes and catenanes (Mabesoone et al. 2020). The self-assembly process often is performed by dissolving the building blocks in a good solvent, then self-assembly is induced by adding a poor solvent that induces the self-assembly of the molecules. The stability and the morphology of the obtained structures are directly conditioned by the solvent conditions used, for instance the ratio between the good and the poor solvents (Korevaar, Schaefer et al. 2012). The kinetics of the self-assembly can be affected by the dynamics of the solvent medium change between the good and the poor solvent; some processing conditions can lead to aggregates that compete for free molecules with the thermodynamically stable structure. P. Korevaar et al. (Korevaar, Schaefer et al. 2012) have developed a model describing the self-assembly process by addition and dissociation of molecules by introducing in their previous model (Korevaar, George et al. 2012) solvent-dependent rate constants. The influence of the mixing protocol on the kinetics of the self-assembly was studied by observing the rate of destruction of the self-assembled structure after contact with a good solvent.

Solvents used in supra-molecular complexation experience competition between different interactions such as host-guest, host-solvent and guest-solvent interactions. Solvents can then be divided in two categories according to their own structure or self-organization:

1. Non-structured or non-self-organized solvents (e.g., halogenated hydrocarbons, aliphatic, aromatic compounds)

2. Structured or self-organized solvents (e.g., water, alcohols).

In non-structured solvents, dispersive interactions dominate and are often an active component in the process of supra-molecular organization. In structured solvents, such as water, the hydrophobic effect plays a major role. This effect refers to the increase in entropy which drives the entire complexation process and that arises from an increase in entropy generated by the expulsion of water molecules when the complex is formed. This entropy is counter-balanced to a small extent, by the decrease in entropy resulting from the host-guest complex formation. Such solvent effects are greatly observed in host-guests systems where interactions with structured or non-structured solvents greatly influence the geometry of the obtained structure. Several excellent treatments of solvation effects and their effects in supra-molecular chemistry, with interesting developments of host-guest interactions mediated by the solvent, are presented in comprehensive articles and reviews, including the works by K. Kanagaraj et al. (Papadakis and Deligkiozi 2019), M. Rekharsky and Y. Inoue (Kanagaraj et al. 2016). The influence of the solvent on the macroscopic properties of supra-molecular arrangements has been evidenced in some studies, where great effects have been observed by subtle changes in solvent. For example, a change in the bulkiness of aromatic solvent molecules while keeping constant the polarity of the solvent, influences the stability of tubular supra-molecular assemblies. Replacing a cyclic alkane solvent by a linear alkane can change the inner conformation or even the helical sense of self-assembled rods (Rekharsky and Inoue 2012). The effect of the solvent on the rheological properties of supra-molecular polymers has been evidenced by using two different solvents with different hydrogen bonding capability. Counter-intuitive behavior has been observed as a stronger hydrogen bonding led to less viscous solutions, these results were related to solvation effect involving the outer corona of the supra-molecular objects and the solvent, and faster local dynamics when cohesive energy between solvent molecules approaches that of the outer corona of the objects.

Physical organogelation is another phenomenon where solvent properties strongly influence the structures obtained. During organogelation, a low-molecular weight compound called organogelator (LMOG) forms a physical gel where a solvent is entrapped; as physical organogelation involves non-covalent interactions such as hydrogen bonding, van der Waals, π stacking, electrostatic and charge-transfer interactions. Indeed, solvent-solvent and solvent-gelator interactions are crucial for organogel formation and strongly define its final properties; some of these interactions may compete with the self-assembly process. Organogelators can self-assemble in filaments. The thickness of these filaments has been seen to strongly influence the rheological properties of the gel. T. Pinault et al. (Alvarenga et al. 2017) studied the gelation of bis-urea derived molecules in aromatic solvents of similar dielectric constant. The strength of hydrogen bonds and dipole moments, and the difference in viscosity of the obtained gels were attributed to the thickness of the filaments obtained after gelation. This thickness was related to the steric effects of the solvents that would be accommodated inside a tubular structure and that would afford different diameters of tubes. The authors propose control of the bulkiness of the solvent molecule as a means to control the size and thus the rheological properties of the obtained materials. The work of M. Mukai et al. (Pinault et al. 2006) has shown that gelator-solvent chirality matching can determine the self-assembly and gelation efficiency in supra-molecular gelation of aspartame lipid gelators in enantiomeric propylene carbonate. Van Esch et al. (Mukai et al. 2012) have observed that the polarity of the solvent strongly influences the contributions of lipophilic and hydrogen bonding interactions to the overall stabilization of gels. Lipophilic interactions usually provide the major stabilizing contribution in polar and hydrogen-bonding competitive solvents; while in lipophilic solvents, hydrogen bonding is the most important stabilizing interaction.

Physical gels can also be formed by polymers in the presence of specific solvents. In this case, worm-like chain conformations presented by the polymer can occur as a result of the intrinsic properties of the polymer, such as in the case of agarose, or be induced by interactions between the polymer and the solvent, as in the case of stereo-regular polymers. Crystalline structures in gels, both in organogels and in polymer gels, are believed to be obtained when one face of the organogel grows faster than the other two, which can be exacerbated by specific interactions with the solvent. D. Dasgupta and J-M. Guenet (Zweep et al. 2009) have shown that the mechanisms of gelation of polymers and of organogelators are different. Fibrils produced with polymers are obtained because chain-folding is impeded, while fibrillar organogels result from the faster grow rate of one face of the crystal lattice with respect to the other two. The effect of the solvent on these both mechanisms has been studied, and it was found to be the key in both cases, which allows us to believe that adjusting solvent properties allows us to obtain different textures and molecular orders in these materials. V. Čaplar et al. (Dasgupta and Guenet 2013) have synthesized a series of LMOGs consisting of aliphatic acid, amino acid (phenylglycine), and ω-aminoaliphatic acid units. By varying the number of methylene units in the aliphatic and in the ω-aminoaliphatic acid chains, a series of positionally isomeric gelators having different positions of the peptidic hydrogen-bonding unit were obtained. The gelation properties of these LMOG were studied in 20 solvents of different structure and polarity. The authors have shown specific solvation effects of isomeric xylenes in the spontaneous resolution into enantiomeric assemblies, followed by different diastomeric aggregation resulting in the formation of polymorphic assemblies. The authors emphasize the impact of specific solvation effects on the determination of the morphology and the stability of the gel fibers obtained.

Y. Xiao et al. (Xiao et al. 2022) have studied the gelation properties of triarylamine tris-amide (TATA) in two isomers of dichlorobenzene: *ortho*-dichlorobenzene and *meta*-dichlorobenzene: these solvents exhibit the same Hansen solubility parameters but produced organogels with different morphologies, molecular structures and thermodynamic properties. Hybrid materials obtained with TATA and PVC thermo reversible gels were prepared as well, to obtain organogelator-polymer hybrid gels. Clearly, from this study, among others, it can be said that the Hansen parameters approach offers only a part of the information of the solvent-gelator system and does not allow prediction of the properties of the organogels in particular morphology, or, to go further, rheological behavior.

K. Aratsu et al. (Aratsu et al. 2020) have studied the effect of the solvent on the thermodynamic stability of toroidal supra-molecular polymers formed by a barbiturated naphthalene molecule in cyclic and acyclic alkanes. The thermodynamic parameters of the supra-molecular arrangements were studied by temperature-dependent UV-vis absorbance. The authors concluded that only dielectric constants (permittivity) of the solvents influence the thermodynamic stability of the supramolecular polymers studied.

M. F. J. Mabesoone et al. (Mabesoone et al. 2020) have reviewed the existing literature about the effects of solute-solvent interactions in supra-molecular polymers, which are aggregates of repeating monomeric units held together by non-covalent interactions, for example, the polymerization of tubuline into microtubules and the filaments of actin. Solvent properties are the key in stabilizing or destabilizing the bound or unbound monomeric units that form the supra-molecular polymer. The most important factors during these phenomena are cohesive forces and dispersive interactions. The thermodynamics of supra-molecular polymers are influenced by entropic and enthalpic contributions to the free energy of the aggregate, which strongly depend on the solvent properties. Besides these thermodynamic aspects, kinetics is also influenced by solvent properties. For instance, addition of a critical concentration of a good solvent to supra-molecular polymers induces disassembly, equilibration kinetics below and above the critical solvent concentration can be modified and gives rise to phenomena such as reaggregation of freed monomers or solubilization of them. Kinetic traps obtained by adding a poor solvent may be different to kinetic traps obtained by thermic effects. Solvents my induce different supra-molecular assembly pathways as in the case of the polymerization of *N*-phenylalanyl decorated perylene diimides in water with 10% of THF that gives rise to concentric rings of left-handed supra-molecular polymers, while the polymerization of this compound in pure THF yielded long fibers with ring-handed helicity. Chiral solvents may induce helicity on supra-molecular polymers, as in the case of chiral triphenylene derivatives (Ślęczkowski et al. 2021). Water has recently been recognized as a compound that can induce structuring in supra-molecular polymerizations as in the case of pyridinium-based monomers that polymerize into helical ribbons when water is present in organic solvents. Water-co-solvent systems can induce specific structuring during polymerizations, as in the case of bypiridine decorated benzene-1, 3, 5-tricarboxamides which form triple helical superstructures in water-isopropanol mixtures that are strongly correlated to the enthalpy of mixing of these solvents (Gillissen et al. 2014). Solvent-co-solvent systems then become a way to control the morphological and physico-chemical properties of supra-molecular assemblies.

The impact of solvents on chirality of supra-molecular assemblies is generally based on three properties of the solvent:

1) polarity, which often induces intermolecular orientation,

2) chirality, as chiral solvents can transfer chirality to the self-assembly performed within this solvent through a chiral induction effect, and

3) active co-assembly with the building blocks.

An example that can be cited is the study of George et al. (Čaplar et al. 2010) that evidenced the decisive role of chiral solvents in the preferred handedness of the self-assemblies formed by oligo (p-phenylene vinylene) achiral molecules, introducing the possibility to control asymmetric non-covalent synthesis directed by the solvent. An excellent review published by S. Xue et al. (George et al. 2011, 2012) addresses the effect of solvents on the control of supra-molecular chirality. The three mentioned properties of the solvent, relevant for the control of the chirality of supra-molecular assemblies, are discussed with very varied examples and finally, the authors give some ideas for future research in exploiting the solvent effect to control supra-molecular chiral structures in different domains. Evidence of the effect of solvent and temperature on the self-assembly of discotic amphiphilic molecules was reported in the study by M. A. J. Gillissen et al. (Xue et al. 2019) where the morphologies of the obtained supra-molecular self-assemblies, in particular, fibrils, were found to be dependent on the composition of the solvent, for instance, water-isopropanol mixtures. Three regimes were proposed by the authors as a function of isopropanol concentration:

Regime I at low isopropanol content, where long fibrous aggregates with a core-shell cross-section were obtained;

Regime II, in the intermediate isopropanol content regime, where objects with low aspect ratio with a core-shell section were found; and

Regime III, at high concentrations of isopropanol, resulting in long fibrous aggregates with a cross-section with single, homogenous contrast.

Excess enthalpy of mixing (ΔH_{mix}^{E}) between water and isopropanol was related to supra-molecular self-assembly observations, with a remarkable correlation between the regime observed with isopropanol concentration and ΔH_{mix}^{E}. The effect of other alcohols was studied too, emphasizing the anomalous behavior of mixtures of alcohol and water, and their effect on supra-molecular assemblies. The effect of solvent composition has also been studied in the self-assembly of an amphiphilic coronene bisimide in THF-water mixtures by K. Venkata and S. J. George (Gillissen et al. 2014).

One-dimensional nanotapes and nanotubes were obtained through π-π stacking and solvophobic interactions; solvent composition was observed to control the morphology and optical properties of the self-assemblies obtained in THF-water mixtures of different concentrations.

It is worthwhile to note here the efforts of several research groups to model the different effects of solvents on supra-molecular systems, in order to

1) understand and predict these effects, and

2) pursue the final objective of fine control of the supra-molecular system arrangement and properties.

A review about molecular modeling of one-dimensional supra-molecular polymers, the mechanism of self-assembly of monomers in polar and non-polar solvents, and the properties they exhibit has been published by D. B. Korlepara and S. Balasubramanian (Rao and George 2010). It presents a good treatment of multi-scale modeling of supra-molecular polymers, and offers good examples of molecular dynamics simulations of supra-molecular systems in non-polar and in polar solvents. Concerning gelation of organic solvents, *a priori* design of organogelators has been proven to be challenging as no organogelator can gel all the solvents. Analysis of solvent parameters and gelator structures can give an understanding of the phenomena occurring in specific systems. In particular, the role of solvent chemistry has been addressed through solvent properties such as polarity, dielectric constant, refractive index, octanol/water partition coefficients, Henry's law constants, and Hildebrand solubility parameters. Multi-term solvent parameters have been used as well to define accurately solvent properties and go from Kamlet-Taft, Catalan, Swain and solvathochromic parameters to Hansen, Flory-Huggins and Modified Separation of Cohesive Energy Density Model (MOSCED) thermodynamic models. An excellent review about these parameters and modeling in the field of gelation has been proposed by Y. Lan (Korlepara and Balasubramanian 2018). In particular, a good treatment of the MOSCED model is presented, which is a model that improves the performance, that is in general poor, of Hansen parameters for associated and solvating systems. The MOSCED model is a thermodynamic one allowing the calculation of the limiting activity coefficients (γ^{∞}) using only pure component parameters. Several versions have been proposed and the latest include binary systems of low polarity, polar and hydrogen bonding compounds.

To describe the energy of interaction of a solute in solution in the MOSCED model, five parameters are proposed: the dispersion parameter λ, the polarity parameter τ, the induction parameter q, the hydrogen-bonding acidity parameter α, and the hydrogen-bonding basicity parameter β. Expressions describing the dependence of τ, α and β with temperature have been developed as well; these MOSCED parameters have been estimated from COSMO-SAC database by M. Gnap and J. Richard Elliot (Lan et al. 2015). Other improvements of MOSCED parameters for solubility of solids (Gnap and Elliott 2018) and for non-electrolytes (Lazzaroni et al. 2005) have been proposed.

Dimethylsulfoxide/Glycerol mixture has been proposed as a model mixture to observe the effects of viscosity of 'friction dependent' chemical reactions by G. Angulo et al. (Angulo et al. 2016) through the observation of the effects on diffusion on chemical reactions. Concentration of glycerol was observed to influence the hydrogen bonding of DMSO allowing control of this property, and to influence chemical reactions that are strongly dependent on viscosity and on hydrogen bonding.

Other oeuvres that can be consulted in relationship with solvent effects in supra-molecular systems topic are the chapter about solvation effects in supra-molecular recognition with an analysis of the Grunwald theory of the compensatory enthalpy–entropy relationship by M. Rekharsky and Y. Inoue (Phifer et al. 2017), and the review about molecular dynamics-based methods for the modeling of soft supra-molecular nanostructures published by T. F. Cova et al. (Rekharsky and Inoue 2012) with a part devoted to thermodynamic calculations of host-guest binding by using molecular dynamics methods for supra-molecular recognition. Recent efforts to understand the role of solvents in dominating the assemblies, structures and properties of coordination supra-molecular systems have been thoroughly reviewed by C. Peng Li and M. Du (Cova 2018). In this review, three scenarios for the solvent are addressed: as a ligand, as a guest, and as both a ligand and a guest in coordination supra-molecular systems. In the case of a solvent being a guest, steric, hydrolysis and synergy effects are discussed. The solvent may induce polymorphism too in crystals by inducing preferential orientation of the molecules and the formation of polymorphs. Properties of coordination supra-molecular systems obtained, such as gas storage and separation capacities, and catalytic, magnetic and optical properties, are affected by the solvents used.

4.7 Dissolution of Minerals

Dissolution of minerals involves molecular scale mechanisms. Minerals dissolution kinetics may impact phenomena as varied as petroleum reservoir stability, carbon sequestration, soil formation, and so on. The study and prediction of mineral dissolution have been the object of several studies (Brand and Bullard 2017; Brand et al. 2017; Juilland and Gallucci 2015; Juilland et al. 2010; Lasaga and Blum 1986). P. Martin et al. (Martin et al. 2020) have proposed a Kinetic Monte Carlo code, called KIMERA (Kinetic Monte Carlo for Mineral Dissolution), to study the evolution in time of a mineral during dissolution, allowing the obtaining of dissolution rates and information about the mechanisms that are involved during dissolution. The program includes information about topographic defects of the crystal such as impurities, dislocations and vacancies, among others. The Arrhenius equation is used in this code to describe the events happening during the dissolution of the crystal, by taking into consideration decoupled phenomena of dissolution and precipitation. It does not, however, explicitly describe the solvent properties and whether mixtures of solvents could be taken into consideration during simulations using KIMERA.

CHAPTER 5

Artificial Intelligence Approaches to Perform Solvent Engineering

The nascent field of Artificial Intelligence and in particular of Machine Learning has been used very recently for applications in the field of solubility in order to tackle problems principally linked to the choice of solvents for industrial or academic applications such as in polymers solubilization, drug development, synthetic routes and chemical process design, extraction and crystallization. For example, S. Boobier et al. (C.-P. Li and Du 2011) have reported an approach combining machine learning and computational chemistry for the prediction of solubility in organic solvents and water. In their approach, a set of 14 descriptors were selected to describe the dissolution phenomenon including the sum of thermal and electronic energies of the solute molecule, solvation energy, orbital interaction between solute and solvent, dipole moment and charge distribution in the solute molecule, molecular volume, Solvent Accessible Surface Area, molecular weight and the number of atoms of the solute. Finally, the experimental melting point was included as a reflection of the sublimation energy of the solid form of the solute. Eight machine learning methods were used in this study, i.e., Multiple Linear Regression (MLR), Partial Least Square (PLS), Artificial Neural Network (ANN), Support Vector Machine (SVM), Gaussian Process (GP), Random Forest (RF), Extra Trees (ET) and Bagging (Bag). This approach gave significantly more accurate predictions compared to other open-access and commercial tools, achieving accuracy close to the expected level of noise in training data (LogS $\pm$ 0.7). It is to be noted that the models used reproduced physico-chemical relationships between solubility and molecular properties in different solvents, leading to rational approaches to improve the accuracy of each model.

In another work, A. Chandrasekaran et al. (Boobier et al. 2020) have presented an approach of Deep Learning for the selection of solvents for polymers. In this work, a data set of 4595 polymers, with their corresponding

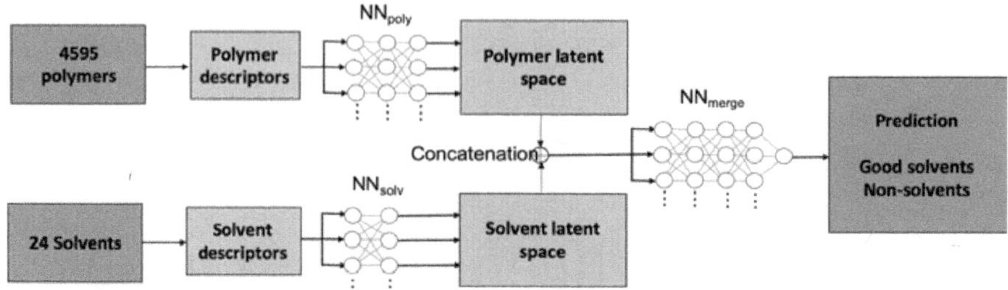

Figure 17. Neural network architecture for the prediction of good solvents and non-solvents for polymers, adapted from A. Chandrasekaran et al. (Chandrasekaran et al. 2020).

solvents and non-solvents was used to train a deep neural network tool taking as input the polymer and solvents descriptors and as output whether a solvent or a non-solvent is good for that polymer. Figure 17 shows the neural network architecture used in this work for the prediction of good solvents and non-solvents for the polymers studied. The proposed method predicted good solvents and non-solvents with an accuracy of 99.6%, which is higher than the accuracy obtained with the Hildebrand and Hansen criteria (60 to 76% accuracy).

Machine Learning techniques have been used to predict gelation of water and/or pure solvents by gelators as in the work presented by F. Delbecq et al. (Chandrasekaran et al. 2020) where the Hansen solubility parameters were used as descriptors for solvents and the output data is the gelation ability represented by a 'gel', 'insoluble' or 'soluble' result. The authors have compared the performances of several methods: nearest neighbors (k-NN), Naïve Bayes, Random Forest, Logistic Regression, Neural Network, Support Vector Machine (SVM), Decision Tree, and Gradient Boosted Tree. The ML methods used performed similarly as predictive tools in this study.

R. Van Lommel et al. (Delbecq et al. 2020) proposed an interesting approach by extracting descriptors to predict gelation by ML from Molecular Dynamics techniques. The four proposed descriptors are: relative solvent accessible surface area, relative end-to-end distance, hydrogen bonding percentage, and the shape factor. These descriptors were obtained for each gelator-solvent pair. The Decision Tree (DT) and the Artificial Neural Networks (ANN) models were used in this study, showing that ANN outperformed the DT model presenting excellent predictive performances.

Other studies have used ML techniques to predict *in silico* gelators able to gel determined solvents and its properties. For instance the work by J. K. Gupta et al. (Van Lommel et al. 2020) used descriptors from QSPR modeling, and then ML models. In this study, three models were revealed to meet criteria to be classified as 'good' models: random forest, support vector machine and neural network. The work by F. Li et al. (Gupta et al.

2016) proposes to design chemical structures that can form hydrogels through combinatorial chemistry and machine learning techniques. On the gel side, mechanical properties were introduced into the models by measuring the increase of G′ (elastic modulus) and G″ (viscous modulus).

The SUSSOL (Sustainable Solvents Selection and Substitution) software has been developed by H. Sels et al. (Sels et al. 2020). Their approach uses artificial intelligence for the selection and substitution of solvents intended to be greener than currently used solvents. The SUSSOL software was used for

i) substitution of existing solvents by other greener candidates and

ii) as a benchmark for new solvents.

This approach is interesting, especially for non-experts in AI and in the solvents domain; however, no mixtures of solvents are included in the treatment of this software. A recent work presented by A. Bakhtyari et al. (Bakhtyari et al. 2023) concerns the application of ML to the modeling of solubility of 19 sugar alcohols in 21 ILs. In this work, adaptive neuro-fuzzy inference system was the best model. Recent approaches such as the physics informed neural networks or physical models embedded in ML (Jirasek and Hasse 2023) have been the object of some studies in domains other than solubility (Peng et al. 2020). However, they are promising routes that may allow an understanding of the complexity of solubility, the role of solvents, and the performance of Solvent Engineering in a straightforward manner.

These recent approaches are or may be the starting point of the massive usage of ML and in general of AI in Solvent Engineering. It is clear that these techniques can be extended to the prediction of the behavior of solutes and/or of gelators, among other molecules, in mixtures of solvents and even on switchable hydrophilicity solvents. However, work is still needed as mixtures of solvents have not yet been addressed.

CHAPTER 6
Concluding Remarks

Throughout this document, we have attempted to show that Solvent Engineering is a powerful tool that can be applied in chemical engineering, processes engineering, physico-chemistry, chemistry and materials science, among other fields. We believe that humanity has arrived at a comprehension of solvents and of solvency phenomena that allows us to engineer these phenomena to adapt them to industrial and, more importantly, to societal needs. Even if the study and comprehension of fine phenomena requires still more thorough work from researchers, this research can be performed in parallel to the development of applications of Solvent Engineering that will allow us to take advantage of the fascinating opportunities that solvents can afford.

References

Abbott, Andrew P., Eric G. Hope, Reena Mistry and Alison M. Stuart. 2009. Probing the structure of gas expanded liquids using relative permittivity, density and polarity measurements. *Green Chemistry* 11(10): 1530–35. http://dx.doi.org/10.1039/B915570H.

Affleck, R., C.A. Haynes and D.S. Clark. 1992. Solvent dielectric effects on protein dynamics. *Proceedings of the National Academy of Sciences* 89(11): 5167 LP–5170. http://www.pnas.org/content/89/11/5167.abstract.

Alfonsi Kim, Colberg Juan, Dunn PeterJ, Fevig Thomas, Jennings Sandra, Johnson Timothy A. et al. 2008. Green chemistry tools to influence a medicinal chemistry and research chemistry based organisation. *Green Chemistry* 10(1): 31–36. http://dx.doi.org/10.1039/B711717E.

Alhashim Salma H., Bhattacharyya Sohini, Tromer Raphael, Kabbani Ahmad, Babu Ganguli, Fernando Oliveira Eliezer et al. 2023. Mechanistic study of lithium-ion battery cathode recycling using deep eutectic solvents. *ACS Sustainable Chemistry & Engineering* 11(18): 6914–22. https://doi.org/10.1021/acssuschemeng.2c06571.

Van Alsten, J.G., C. Hansen and C.A. Eckert. 1984. Supercritical enhancement factors for nonpolar and polar systems. *In: AIChE Annual Meeting.* Paper 84a.

Alvarenga, Bruno G., Matthieu Raynal, Laurent Bouteiller and Edvaldo Sabadini. 2017. Unexpected solvent influence on the rheology of supramolecular polymers. *Macromolecules* 50(17): 6631–36. https://doi.org/10.1021/acs.macromol.7b00786.

Amar Yehia, Schweidtmann Artur M., Deutsch Paul, Cao Liwei and Lapkin Alexei. 2019. Machine learning and molecular descriptors enable rational solvent selection in asymmetric catalysis. *Chemical Science* 10(27): 6697–6706. http://dx.doi.org/10.1039/C9SC01844A.

Amenta Valeria, Cook Joanne L., Hunter Christopher A., Low Caroline M. R., Sun Hongmei and Vinter Jeremy G. 2013. Interplay of self-association and solvation in polar liquids. *Journal of the American Chemical Society* 135(32): 12091–100. https://doi.org/10.1021/ja405799q.

Anastas, Paul and Nicolas Eghbali. 2010. Green chemistry: principles and practice. *Chemical Society Reviews* 39(1): 301–12.

Angulo Gonzalo, Brucka Marta, Gerecke MArio, Grampp Günter, Jeannerat Damien, Milkiewicz Jadwiga et al. 2016. Characterization of dimethylsulfoxide/glycerol mixtures: a binary solvent system for the study of 'friction-dependent' chemical reactivity. *Physical Chemistry Chemical Physics* 18(27): 18460–69. http://dx.doi.org/10.1039/C6CP02997C.

Anugwom Ikenna, Mäki-Arvela Päivi, Virtanen Pasi, Damlin Pia, Sjöholm Rainer and Mikkola Jyri-Pekka. 2011. Switchable Ionic Liquids (SILs) based on glycerol and acid gases. *RSC Advances* 1(3): 452–57. http://dx.doi.org/10.1039/C1RA00154J.

Anqwaar N. Muhammad, Yilgor Emel, Baris YAgci Mustafa, Unal Ugur and Yilgor Iskender. 2019. Effect of reaction solvent on hydroxyapatite synthesis in sol–gel process. *Royal Society Open Science* 4(12): 171098. https://doi.org/10.1098/rsos.171098.

Arain, Zulqarnain, Liu Chen, Yang Yi, Mateen M., Ren Yinken, Ding Yong et al. 2019. Elucidating the dynamics of solvent engineering for perovskite solar cells. *Science China Materials* 62(2): 161–72. http://engine.scichina.com/doi/10.1007/s40843-018-9336-1.

Araque, Juan C., Jeevapani J. Hettige and Claudio J. Margulis. 2015. Ionic liquids—conventional solvent mixtures, structurally different but dynamically similar. *The Journal of Chemical Physics* 143(13): 134505. https://doi.org/10.1063/1.4932331.

Aratsu Keisuke, Shimizu Nobutaka, Takagi Hideaki, Haruki Rie, Adachi Shin-ichi and Yagai Shiki. 2020. Effect of solvent on the thermodynamic stability of toroidal supramolecular polymers. *Chemistry Letters* 49(2): 178–81. https://cir.nii.ac.jp/crid/1390002184872315392.

Augustijns, Patrick and Marcus E. Brewster (eds.). 2007. *Solvent Systems and Their Selection in Pharmaceutics and Biopharmaceutics*. Springer New York. https://doi.org/10.1007/978-0-387-69154-1.

Backes Sebastian, Krause Patrick, Tabaka Weronika, Witt Marcus U., Mukherji Debashish, Kremer Kurt et al. 2017. Poly(N-Isopropylacrylamide) microgels under alcoholic intoxication: when a LCST polymer shows swelling with increasing temperature. *ACS Macro Letters* 6(10): 1042–46. https://doi.org/10.1021/acsmacrolett.7b00557.

Bahadori, Alireza, Hari B. Vuthaluru and Saeid Mokhatab. 2009. New correlations predict aqueous solubility and density of carbon dioxide. *International Journal of Greenhouse Gas Control* 3(4): 474–80. http://www.sciencedirect.com/science/article/pii/S1750583609000139.

Bajwa, D.S., G. Pourhashem, A.H. Ullah and S.G. Bajwa. 2019. A concise review of current lignin production, applications, products and their environmental impact. *Industrial Crops and Products* 139: 111526. https://www.sciencedirect.com/science/article/pii/S0926669019305382.

Bakhtyari, Ali, Ali Rasoolzadeh, Behzad Vaferi and Amith Khandakar. 2023. Application of machine learning techniques to the modeling of solubility of sugar alcohols in ionic liquids. *Scientific Reports* 13(1): 12161. https://doi.org/10.1038/s41598-023-39441-7.

Ballesteros-Gómez, Ana, Soledad Rubio and Dolores Pérez-Bendito. 2009. Potential of supramolecular solvents for the extraction of contaminants in liquid foods. *Journal of Chromatography A* 1216(3): 530–39. http://www.sciencedirect.com/science/article/pii/S0021967308010455.

Ballesteros-Gómez, Ana, María Dolores Sicilia and Soledad Rubio. 2010. Supramolecular solvents in the extraction of organic compounds. a review. *Analytica Chimica Acta* 677(2): 108–30. http://www.sciencedirect.com/science/article/pii/S0003267010009359.

Banuti, D.T. 2015. Crossing the widom-line – supercritical pseudo-boiling. *The Journal of Supercritical Fluids* 98: 12–16. http://www.sciencedirect.com/science/article/pii/S0896844614004306.

Bär, Markus, Robert Großmann, Sebastian Heidenreich and Fernando Peruani. 2020. Self-propelled rods: insights and perspectives for active matter. *Annual Review of Condensed Matter Physics* 11(1): 441–66. https://doi.org/10.1146/annurev-conmatphys-031119-050611.

Baraban, L., M. Tasinkevych, M.N. Popescu, S. Sanchez, S. Dietrich and O.G. Schmidt. 2012. Transport of cargo by catalytic janus micro-motors. *Soft Matter* 8(1): 48–52. http://dx.doi.org/10.1039/C1SM06512B.

Barton, B.F., J.L. Reeve and A.J. McHugh. 1997. Observations on the dynamics of nonsolvent-induced phase inversion. *Journal of Polymer Science Part B: Polymer Physics* 35(4): 569–85. https://doi.org/10.1002/(SICI)1099-0488(199703)35:4%3C569::AID-POLB5%3E3.0.CO.

Belonoshko, A. and S.K. Saxena. 1991. A molecular dynamics study of the pressure-volume-temperature properties of supercritical fluids: II. CO_2, CH_4, CO, O_2, and H_2. *Geochimica et Cosmochimica Acta* 55(11): 3191–3208. http://www.sciencedirect.com/science/article/pii/001670379190483L.

Benedix Robin R., Botsch Sophia, Preisig Natalie, Kovalchik Volodymyr, Jessop Philip G. and Stubenrauch Cosima. 2023. Influence of a CO_2-switchable additive on the surface and foaming properties of a cationic non-switchable surfactant. *Soft Matter* 19(16): 2941–48. http://dx.doi.org/10.1039/D3SM00273J.

Bennett, T.D., F.X. Coudert, S.L. James and A.I. Cooper. 2021. The changing state of porous materials. *Nat Mater.* 20(9): 1179–87. https://www.nature.com/articles/s41563-021-00957-w#citeas.

Bera, Arabinda, Soudamini Sahoo, Snigdha Thakur and Subir K. Das. 2022. Active particles in explicit solvent: dynamics of clustering for alignment interaction. *Physical Review E* 105(1): 14606. https://link.aps.org/doi/10.1103/PhysRevE.105.014606.

Bernal-García J. Manuel, Guzmán-López Adriana, Cabrales-Torres Alberto, Estrada-Baltazar Alejandro and Iglesias-Silva Gustavo A. 2008. Densities and viscosities of (N,N-dimethylformamide + water) at atmospheric pressure from (283.15 to 353.15) K. *Journal of Chemical & Engineering Data* 53(4): 1024–27. https://doi.org/10.1021/je700671t.

Bertolasi, Valerio, Paola Gilli, Valeria Ferretti and Gastone Gilli. 2001. Competition between hydrogen bonding and donor-acceptor interactions in co-crystals of 1,3-dimethylbarbituric acid with aromatic amines. *New Journal of Chemistry* 25(3): 408–15. http://dx.doi.org/10.1039/B008262G.

Bezerra, Marcos de Almeida, Marco Aurélio Zezzi Arruda and Sérgio Luis Costa Ferreira. 2005. Cloud point extraction as a procedure of separation and pre-concentration for metal determination using spectroanalytical techniques: a review. *Applied Spectroscopy Reviews* 40(4): 269–99. https://doi.org/10.1080/05704920500230880.

Bischofberger, I., D.C.E. Calzolari and V. Trappe. 2014. Co-nonsolvency of PNiPAM at the transition between solvation mechanisms. *Soft Matter* 10(41): 8288–95. http://dx.doi.org/10.1039/C4SM01345J.

Bogel-Łukasik, E., R. Bogel-Łukasik, K. Kriaa, I. Fonseca, Y. Tarasenko and M. Nunes da Ponte. 2008. Limonene hydrogenation in high-pressure CO_2: Effect of hydrogen pressure. *The Journal of Supercritical Fluids* 45(2): 225–30. http://www.sciencedirect.com/science/article/pii/S089684460700397X.

Boobier, Samuel, David R.J. Hose, A. John Blacker and Bao N. Nguyen. 2020. Machine learning with physicochemical relationships: solubility prediction in organic solvents and water. *Nature Communications* 11(1): 5753. https://doi.org/10.1038/s41467-020-19594-z.

Borjani Tohid, GArcia-Munoz Salvador, Luciani Carla Vanesa Galindo Amparo and Adjiman Claire S. 2019. Hybrid QSPR models for the prediction of the free energy of solvation of organic solute/solvent pairs. *Physical Chemistry Chemical Physics* 21(25): 13706–20. http://dx.doi.org/10.1039/C8CP07562J.

Borne, Isaiah, Natalie Simon, Christopher W. Jones and Ryan P. Lively. 2022. Design of gas separation processes using type II porous liquids as physical solvents. *Industrial & Engineering Chemistry Research* 61(32): 11908–21. https://doi.org/10.1021/acs.iecr.2c01943.

Böttcher Norbert, Taron Joshua, Kolditz Olaf, Park Chan-Hee and Liedl Rudolf. 2012. Evaluation of thermal equations of state for CO_2 in numerical simulations. *Environmental Earth Sciences* 67(2): 481–95. https://doi.org/10.1007/s12665-012-1704-1.

Boulton, James R. and Fred P. Stein. 1993. The constant pressure heat capacity of supercritical carbon dioxide-methanol and carbon dioxide-ethanol co-solvent mixtures. *Fluid Phase Equilibria* 91(1): 159–76. http://www.sciencedirect.com/science/article/pii/0378381293850862.

Boyd, Alaina R., Philip G. Jessop, Julian M. Dust and Erwin Buncel. 2013. Switchable polarity solvent (SPS) systems: Probing solvatoswitching with a Spiropyran (SP)–Merocyanine (MC) Photoswitch. *Organic & Biomolecular Chemistry* 11(36): 6047–55. http://dx.doi.org/10.1039/C3OB41204K.

Bradford, E., A.M. Schweidtmann and A. Lapkin. 2018. Efficient multiobjective optimization employing gaussian processes, spectral sampling and a genetic algorithm. *J. Glob Optim* 71(2): 407–38. https://link.springer.com/article/10.1007/s10898-018-0609-2.

Branam, Richard. 2005. Molecular Dynamics Simulation of Supercritical Fluids. Pennsylvania State University.

Brand, Alexander S. and Jeffrey W. Bullard. 2017. Dissolution kinetics of cubic tricalcium aluminate measured by digital holographic microscopy. *Langmuir* 33(38): 9645–56. https://doi.org/10.1021/acs.langmuir.7b02400.

Brand, Alexander S., Pan Feng and Jeffrey W. Bullard. 2017. Calcite dissolution rate spectra measured by *in situ* digital holographic microscopy. *Geochimica et Cosmochimica Acta* 213: 317–29. https://www.sciencedirect.com/science/article/pii/S001670371730409X.

Brunner, G. 2004. *Supercritical Fluids as Solvents and Reaction Media.* Elsevier. https://doi.org/10.1016/B978-0-444-51574-2.X5000-7.

Caballo, Carmen, María D. Sicilia and Soledad Rubio. 2017. Chapter 5 - Supramolecular solvents for green chemistry. pp. 111–37. *In:* Francisco Pena-Pereira and Marek, B.T. (eds.). The Application of Green Solvents in Separation Processes Tobiszewski. Elsevier. http://www.sciencedirect.com/science/article/pii/B978012805297600005X.

Cabot, Rafel and Christopher A. Hunter. 2010. A thermodynamic study of selective solvation in solvent mixtures. *Organic & Biomolecular Chemistry* 8(8): 1943–50. http://dx.doi.org/10.1039/B927107D.

Cameretti, Luca F., Gabriele Sadowski and Jørgen M. Mollerup. 2005. Modeling of aqueous electrolyte solutions with perturbed-chain statistical associated fluid theory. *Industrial & Engineering Chemistry Research* 44(9): 3355–62. https://doi.org/10.1021/ie0488142.

Cao Xia, Gao Peiyuan, Ren Xiaodi and Zhang Ji-Guang. 2021. Effects of fluorinated solvents on electrolyte solvation structures and electrode/electrolyte interphases for lithium metal batteries. *Proceedings of the National Academy of Sciences* 118(9): e2020357118. https://doi.org/10.1073/pnas.2020357118.

Cao Xiaobing, Zhi Lili, Jia Yi, Li Yahui, Zhao Ke, Cui Xian et al. 2019. A review of the role of solvents in formation of high-quality solution-processed perovskite films. *ACS Applied Materials & Interfaces* 11(8): 7639–54. https://doi.org/10.1021/acsami.8b16315.

Cao, Xiaobing, Lei Hao, Gengyang Su, Xiaoxi Li, a Tuyu Dong, a Pengjie Chao, Daize Mo, Qingguang Zeng, Xin He and Jinquan Wei. 2023. Green solvents processed all functional layers for efficient perovskite solar cells. *RSC Sustainability* 1(5): 1290–97. http://dx.doi.org/10.1039/D3SU00146F.

Čaplar, Vesna, Leo Frkanec, Nataša Šijaković Vujičić and Mladen Žinić. 2010. Positionally isomeric organic gelators: structure–gelation study, racemic versus enantiomeric gelators, and solvation effects. *Chemistry – A European Journal* 16(10): 3066–82. https://doi.org/10.1002/chem.200902342.

de Castro, Carlos A. Nieto, Murshed S.M. Sohel, J.V. Lourenço Maria, J.V. Santos Fernando, L.M. Lopes Manuel and M.P. França Joao. 2012. Ionanofluids: new heat transfer fluids for green processes development BT - Green solvents I: Properties and applications in chemistry. pp. 233–49.

In: Ali Mohammad (ed.). Dordrecht: Springer Netherlands. https://doi.org/10.1007/978-94-007-1712-1_8.

Catalán, Javier. 2009. Toward a generalized treatment of the solvent effect based on four empirical scales: dipolarity (SdP, a New Scale), Polarizability (SP), Acidity (SA), and Basicity (SB) of the medium. *The Journal of Physical Chemistry B* 113(17): 5951–60. https://doi.org/10.1021/jp8095727.

Catoni, E., E. Cernia and C. Palocci. 1996. Different aspects of 'solvent engineering' in lipase biocatalysed esterifications. *Journal of Molecular Catalysis A: Chemical* 105(1): 79–86. https://www.sciencedirect.com/science/article/pii/1381116995001530.

Caviglia, G. and A. Morro. 2015. Continuum model of cell motility and chemotaxis. *European Journal of Mechanics - B/Fluids* 49: 235–42. http://www.sciencedirect.com/science/article/pii/S0997754614001472.

Celebre Giorgio, De Luca Giuseppina, Maiorino Michela, Iemma Francesca, Ferrarini Alberta, Pieraccini Silvia et al. 2005. Solute−solvent interactions and chiral induction in liquid crystals. *Journal of the American Chemical Society* 127(33): 11736–44. https://doi.org/10.1021/ja051589a.

Chandrasekaran, Anand, Chiho Kim, Shruti Venkatram and Rampi Ramprasad. 2020. A deep learning solvent-selection paradigm powered by a massive solvent/nonsolvent database for polymers. *Macromolecules* 53(12): 4764–69. https://doi.org/10.1021/acs.macromol.0c00251.

Chang Chao-Wen, Borne Isaiah, Lawler Robin M., Yu Zhenzi, Jang Seung Soon, Lively Ryan P. et al. 2022. Accelerating solvent selection for Type II porous liquids. *Journal of the American Chemical Society* 144(9): 4071–79. https://doi.org/10.1021/jacs.1c13049.

Chapman, W.G., K.E. Gubbins, G. Jackson and M. Radosz. 1989. SAFT: Equation-of-state solution model for associating fluids. *Fluid Phase Equilibria* 52: 31–38. https://www.sciencedirect.com/science/article/pii/0378381289803085.

Clark, J.H., A. Hunt, C. Topi, G. Paggiola and J. Sherwood. 2017. CHAPTER 1 Introduction to solvents and sustainable chemistry. pp. 1–34. *In: Sustainable Solvents: Perspectives from Research, Business and International Policy*, The Royal Society of Chemistry. https://books.rsc.org/books/monograph/633/chapter/315174/Introduction-to-Solvents-and-Sustainable-Chemistry.

Chen, Geping and Yigang Lu. 2015. Cavitation in compressible supercritical carbon dioxide. *Physics and Chemistry of Liquids* 53(1): 67–74. https://doi.org/10.1080/00319104.2014.915711.

Chen Yu, Liu Chong, Wang Yanlong, Tian Yurun, Li Yuting, Feng Minghui et al. 2023. Efficient recovery of valuable metals from lithium-ion battery cathodes using phytic acid-based deep eutectic solvents at a mild temperature. *Energy & Fuels* 37(7): 5361–69. https://doi.org/10.1021/acs.energyfuels.3c00313.

Cheng Jiajie, Fan Zhenjun and Dong Jingjing. 2023. Research progress of green solvent in CsPbBr3 perovskite solar cells. *Nanomaterials (Basel, Switzerland)* 13(6): 991.

Cho Jaemin, Kim Beomsoo, Ryu Seokjoo, Yun Alan Jiwan, Gil Bumjin, Lim Jiheon et al. 2023. Multifunctional green solvent for efficient perovskite solar cells. *Electronic Materials Letters* 19(5): 462–70. https://doi.org/10.1007/s13391-023-00410-x.

Choa Jung Sang, Na Koa You, Kooa Hye Young, Leeb Man-Jong and Kanga Youn Chan. 2009. Effects of solvent on the properties of nano-sized hydroxyapatite powders directly prepared by high temperature flame spray pyrolysis. *Journal of Ceramic Processing Research* 10: 628–32.

Christian Reichardt and Thomas Welton. 2010. *Solvents and Solvent Effects in Organic Chemistry*. ed. Wiley-VCH Verlag GmbH & Co. KGaA.

Chuang, W.-Y., T.-H. Young, W.-Y Chiu and C.-Y. Lin. 2000. The effect of polymeric additives on the structure and permeability of poly(vinyl alcohol) asymmetric membranes. *Polymer* 41(15): 5633–41. http://www.sciencedirect.com/science/article/pii/S0032386199008186.

Clarke Coby J., Tu Wei-Chien, Levers Olivier, Bröhl Andreas and Hallett Jason P. 2018. Green and sustainable solvents in chemical processes. *Chemical Reviews* 118(2): 747–800. https://doi.org/10.1021/acs.chemrev.7b00571.

Claverie Marie, Martin François, Careme Christel, Le Roux Christophe, Micoud Pierre, Grauby Olivier et al. 2018. Flow supercritical synthesis of brucite and magnesian T-O, T-O-T Phyllosilicates: An opportunity to tune the structure with the solvent composition. *Clay Minerals* 53(3): 497–503. https://www.cambridge.org/core/article/flow-supercritical-synthesis-of-brucite-and-magnesian-to-tot-phyllosilicates-an-opportunity-to-tune-the-structure-with-the-solvent-composition/9A7C62FDE8B7A95E62C3EBF783598812.

Cohen, Claude, G.B. Tanny and Stephen Prager. 1979. Diffusion-controlled formation of porous structures in ternary polymer systems. *Journal of Polymer Science: Polymer Physics Edition* 17(3): 477–89. https://doi.org/10.1002/pol.1979.180170312.

Colina, C.M., C.G. Olivera-Fuentes, F.R. Siperstein, M. Lisal and K.E. Gubbins. 2003. Thermal properties of supercritical carbon dioxide by monte carlo simulations. *Molecular Simulation* 29(6–7): 405–12. https://doi.org/10.1080/0892702031000117135.

Colja, Laane, Boeren Sjef, Vos Kees and Veeger Cees. 2018. Rules for optimization of biocatalysis in organic solvents. *Biotechnology and Bioengineering* 30(1): 81–87. https://doi.org/10.1002/bit.260300112.

Cook Joanne L., Hunter Christopher A., M.R. Low Caroline, Perez-Velasco Alejandro and Vinter Jeremy G. 2007. Solvent effects on hydrogen bonding. *Angewandte Chemie International Edition* 46(20): 3706–9. https://doi.org/10.1002/anie.200604966.

Joanne L. Cook, Christopher A. Hunter, Caroline M. R. Low, Alejandro Perez-Velasco and Jeremy G. Vinter. 2008. Preferential solvation and hydrogen bonding in mixed solvents. *Angewandte Chemie International Edition* 47(33): 6275–77. https://doi.org/10.1002/anie.200801349.

Cova, Tânia F. 2018. Modeling Soft Supramolecular Nanostructures by Molecular Simulations. *In*: Sandra C. Nunes (ed.). Rijeka: IntechOpen, Ch. 2. https://doi.org/10.5772/intechopen.74939.

Cramer, M.S. 2012. Numerical estimates for the bulk viscosity of ideal gases. *Physics of Fluids* 24(6): 66102. https://doi.org/10.1063/1.4729611.

da Cruz Márcia G.A., Gueret Robin, Chen Jianhong, Piątek Jędrzej, Beele Björn, Sipponen Björn et al. 2022. Electrochemical depolymerization of lignin in a biomass-based solvent**. *ChemSusChem* 15(15): e202200718. https://doi.org/10.1002/cssc.202200718.

da Cruz, M.G.A., B.V.M. Rodrigues, A. Ristic, S. Budnyk, S. Das and A. Slabon. 2022. On the product selectivity in the electrochemical reductive cleavage of 2-phenoxyacetophenone, a lignin model compound. *Green Chemistry Letters and Reviews* 15(1): 153–61. https://doi.org/10.1080/17518253.2022.2025462.

Curran, S.A., D. Zhang, W.T. Wondmagegn, A.V. Ellis, J. Cech, S. Roth et al. 2006. Dynamic electrical properties of polymer-carbon nanotube composites: enhancement through covalent bonding. *Journal of Materials Research* 21(4): 1071–77. https://www.cambridge.org/core/article/dynamic-electrical-properties-of-polymercarbon-nanotube-composites-enhancement-through-covalent-bonding/5F46B28675559D7BC76E2EE812FBF4BE.

Dasgupta, Debarshi and Jean-Michel Guenet. 2013. The solvent in physical gelation: polymers versus organogelators. *Macromolecular Chemistry and Physics* 214(17): 1885–92. https://doi.org/10.1002/macp.201300094.

Day, Chany-Yih, Chiehming J. Chang and Chiu-Yang Chen. 1999. Phase equilibrium of Ethanol + CO_2 and Acetone + CO_2 at elevated pressures. *Journal of Chemical & Engineering Data* 44(2): 365. https://doi.org/10.1021/je9804890.

Delbecq, Frederic, Guillaume Adenier, Yuki Ogue and Takeshi Kawai. 2020. Gelation properties of various long chain amidoamines: prediction of solvent gelation via machine learning using hansen solubility parameters. *Journal of Molecular Liquids* 303: 112587. https://www.sciencedirect.com/science/article/pii/S0167732219351967.

Deng Zheng, Ying Wen, Gong Ke, Zeng Yu-Jia, Yan Youguo and Peng Xinsheng. 2020. Facilitate gas transport through metal-organic polyhedra constructed porous liquid membrane. *Small* 16(11): 1907016. https://doi.org/10.1002/smll.201907016.

Devi Tulsiyan, Kiran, Mallika Rani Prusty and Himansu S. Biswal. 2023. Effect of choline amino acid-based ionic liquids on stability and structure of hemoglobin. *ChemPhysChem* 24(15): e202300201. https://doi.org/10.1002/cphc.202300201.

Dewar, Michael J.S. and J.P. Schroeder. 1964. Liquid crystals as solvents. I. The use of nematic and smectic phases in gas-liquid chromatography. *Journal of the American Chemical Society* 86(23): 5235–39. https://doi.org/10.1021/ja01077a039.

Ding Michael S., Li Qiuyan, Li Xing, Xu Wu and Xu Kang. 2017. Effects of solvent composition on liquid range, glass transition, and conductivity of electrolytes of a (Li, Cs)PF6 salt in EC-PC-EMC solvents. *The Journal of Physical Chemistry C* 121(21): 11178–83. https://doi.org/10.1021/acs.jpcc.7b03306.

Domínguez de María, Pablo. 2014. Recent trends in (ligno)cellulose dissolution using neoteric solvents: switchable, distillable and bio-based ionic liquids. *Journal of Chemical Technology & Biotechnology* 89(1): 11–18. https://doi.org/10.1002/jctb.4201.

Drago, Russell S. 1992. A unified scale for understanding and predicting non-specific solvent polarity: a dynamic cavity model. *Journal of the Chemical Society, Perkin Transactions* 2(10): 1827–38. http://dx.doi.org/10.1039/P29920001827.

Driver, Mark D., Mark J. Williamson, Joanne L. Cook and Christopher A. Hunter. 2020. Functional group interaction profiles: a general treatment of solvent effects on non-covalent interactions. *Chemical Science* 11(17): 4456–66. http://dx.doi.org/10.1039/D0SC01288B.

Duan, Zhenhao, Nancy Møller and John H. Weare. 1996. A general equation of state for supercritical fluid mixtures and molecular dynamics simulation of mixture PVTX properties. *Geochimica et Cosmochimica Acta* 60(7): 1209–16. http://www.sciencedirect.com/science/article/pii/001670379600004X.

Dudowicz, Jacek, Karl F. Freed and Jack F. Douglas. 2015. Communication: cosolvency and cononsolvency explained in terms of a flory-huggins type theory. *The Journal of Chemical Physics* 143(13): 131101. https://aip.scitation.org/doi/abs/10.1063/1.4932061.

Ebbens, S.J. 2016. Active colloids: progress and challenges towards realising autonomous applications. *Current Opinion in Colloid & Interface Science* 21: 14–23. http://www.sciencedirect.com/science/article/pii/S1359029415000679.

Eckert, Charles A., Barbara L. Knutson and Pablo G. Debenedetti. 1996. Supercritical fluids as solvents for chemical and materials processing. *Nature* 383(6598): 313–18. https://doi.org/10.1038/383313a0.

Emanuel, George. 1992. Effect of bulk viscosity on a hypersonic boundary layer. *Physics of Fluids A: Fluid Dynamics* 4(3): 491–95. https://doi.org/10.1063/1.858322.

Fanali, Salvatore, Paul R. Haddad, Colin Poole and Marja-Liisa Riekkola (eds.). 2017. *Liquid Chromatography. Fundamentals and Instrumentation*. Elsevier. https://doi.org/10.1016/C2015-0-04294-0.

Fields, P.R., T.L. Chester and A.M. Stalcup. 2011. Viscosity estimation in binary and ternary supercritical fluid mixtures containing carbon dioxide using a supercritical fluid chromatograph. *Journal of Liquid Chromatography & Related Technologies* 34(12): 995–1003. https://doi.org/10.1080/10826076.2011.565540.

Figoli, A., T. Marino, S. Simone, Nicolo E. Di, X-M. Li, T. He et al. 2014. Towards non-toxic solvents for membrane preparation: a review. *Green Chemistry* 16(9): 4034–59. http://dx.doi.org/10.1039/C4GC00613E.

Folić, Milica, Claire S. Adjiman and Efstratios N. Pistikopoulos. 2008. Computer-aided solvent design for reactions: maximizing product formation. *Industrial & Engineering Chemistry Research* 47(15): 5190–5202. https://doi.org/10.1021/ie0714549.

Fomin, Yu. D., V.N. Ryzhov, E.N. Tsiok and V.V. Brazhkin. 2015. Thermodynamic properties of supercritical carbon dioxide: widom and frenkel lines. *Physical Review E* 91(2): 22111. https://link.aps.org/doi/10.1103/PhysRevE.91.022111.

Ford, Jackson W., Jie Lu, Charles L. Liotta and Charles A. Eckert. 2008. Solvent effects on the kinetics of a diels–alder reaction in gas-expanded liquids. *Industrial & Engineering Chemistry Research* 47(3): 632–37. http://pubs.acs.org/doi/abs/10.1021/ie070618i.

Freedman, Miriam Arak. 2020. Liquid–liquid phase separation in supermicrometer and submicrometer aerosol particles. *Accounts of Chemical Research* 53(6): 1102–10. https://doi.org/10.1021/acs. accounts.0c00093.

Friend, D.G. and H.M. Roder. 1987. The thermal conductivity surface for mixtures of methane and ethane. *International Journal of Thermophysics* 8(1): 13–26. https://doi.org/10.1007/ BF00503221.

Friend, Daniel G. and Hans M. Roder. 1985. Thermal-conductivity enhancement near the liquid-vapor critical line of binary methane-ethane mixtures. *Physical Review A* 32(3): 1941–44. https://link. aps.org/doi/10.1103/PhysRevA.32.1941.

Fukunaga, Kenji and Takehisa Kunieda. 1999. Liquid crystal control. a remarkable enhancement of both efficiency and diastereoselectivity of intramolecular thermal cycloadditions in smectic solvents. *Tetrahedron Letters* 40(33): 6041–44. https://www.sciencedirect.com/science/article/ pii/S0040403999012009.

Galliéro, Guillaume, Christian Boned and Antoine Baylaucq. 2005. Molecular dynamics study of the lennard–jones fluid viscosity: application to real fluids. *Industrial & Engineering Chemistry Research* 44(17): 6963–72. https://doi.org/10.1021/ie050154t.

Gekko, Kunihiko and Hiroharu Ito. 1990. Competing solvent effects of polyols and guanidine hydrochloride on protein stability. *The Journal of Biochemistry* 107(4): 572–77. https://doi. org/10.1093/oxfordjournals.jbchem.a123088.

George Subi J., de Bruijin Robin, Tomovic Zeljko, Van Averbeke Bernard, Beljonne David, Lazzaroni Roberto et al. 2012. Asymmetric noncovalent synthesis of self-assembled one-dimensional stacks by a chiral supramolecular auxiliary approach. *Journal of the American Chemical Society* 134(42): 17789–96. https://doi.org/10.1021/ja3086005.

George, Subi J., Željko Tomović, Albertus P.H.J. Schenning and E.W. Meijer. 2011. Insight into the chiral induction in supramolecular stacks through preferential chiral solvation. *Chemical Communications* 47(12): 3451–53. http://dx.doi.org/10.1039/C0CC04617E.

Gerhards Tina, Mackfeld Ursula, Bocola Marco, von Lieres Eric, Wiechert Wolfang, Pohl Martina et al. 2012. Influence of organic solvents on enzymatic asymmetric carboligations. *Advanced Synthesis & Catalysis* 354(14–15): 2805–20. https://pubmed.ncbi.nlm.nih.gov/23349644.

Ghosh Pradyut, Federwisch Guido, Kogej Michael, Scalley Christoph A., Haase Detlev, Saak Wolfang et al. 2005. Controlling the rate of shuttling motions in [2]rotaxanes by electrostatic interactions: a cation as solvent-tunable brake. *Organic & Biomolecular Chemistry* 3(15): 2691–2700. http://dx.doi.org/10.1039/B506756A.

Gillissen Martijn A.J., M.E. Koenigs Marcel, J.H. Spiering Jolanda, A.J.M. Vekemans Jef, R.A. Palmans Anja, K. Voets Ilja et al. 2014. Triple helix formation in amphiphilic discotics: demystifying solvent effects in supramolecular self-assembly. *Journal of the American Chemical Society* 136(1): 336–43. https://doi.org/10.1021/ja4104183.

Giomi, Luca, Tanniemola B. Liverpool and M. Cristina Marchetti. 2010. Sheared active fluids: thickening, thinning, and vanishing viscosity. *Physical Review E* 81(5): 51908. https://link.aps. org/doi/10.1103/PhysRevE.81.051908.

Gnap, Marshall and J. Richard Elliott. 2018. Estimation of MOSCED parameters from the COSMO-SAC database. *Fluid Phase Equilibria* 470: 241–48. http://www.sciencedirect.com/science/ article/pii/S0378381218300402.

Gonnella, Giuseppe, Davide Marenduzzo, Antonio Suma and Adriano Tiribocchi. 2015. Motility-induced phase separation and coarsening in active matter. *Comptes Rendus Physique* 16(3): 316–31. http://www.sciencedirect.com/science/article/pii/S1631070515000791.

González, Begoña, Noelia Calvar, Elena Gómez and Ángeles Domínguez. 2007. Density, dynamic viscosity, and derived properties of binary mixtures of methanol or ethanol with water, ethyl acetate, and methyl acetate at T = (293.15, 298.15, and 303.15)K. *The Journal of Chemical*

Thermodynamics 39(12): 1578–88. http://www.sciencedirect.com/science/article/pii/S0021961407000900.

Gonzalez, H. and G. Emanuel. 1993. Effect of bulk viscosity on couette flow. *Physics of Fluids A: Fluid Dynamics* 5(5): 1267–68. https://doi.org/10.1063/1.858612.

Goodyear, Grant, Michael W. Maddox and Susan C. Tucker. 2000. Domain-based characterization of density inhomogeneities in compressible supercritical fluids. *The Journal of Physical Chemistry B* 104(26): 6240–47. https://doi.org/10.1021/jp000378j.

Goossens, Karel, Kathleen Lava, Christopher W. Bielawski and Koen Binnemans. 2016. Ionic liquid crystals: versatile materials. *Chemical Reviews* 116(8): 4643–4807. https://doi.org/10.1021/cr400334b.

Granero-Fernandez Emanuel, Machin Devin, Lacaze-Dufaure Corinne, Camy Séverine, Condoret Jean-Stéphane, Gerbaud Vincent et al. 2018. CO_2-expanded alkyl acetates: physicochemical and molecular modeling study and applications in chemical processes. *ACS Sustainable Chemistry & Engineering* 6(6): 7627–37. https://doi.org/10.1021/acssuschemeng.8b00454.

Granero Fernandez, Emanuel, Jean-Stéphane Condoret, Vincent Gerbaud and Yaocihuatl Medina-Gonzalez. 2017. Molecular dynamics simulations of gas-expanded liquids. pp. 175–80. *In*: Antonio Espuña, Moisès Graells and Luis B.T. (eds.). *27 European Symposium on Computer Aided Process Engineering* - Computer Aided Chemical Engineering Puigjaner. Elsevier. http://www.sciencedirect.com/science/article/pii/B9780444639653500313.

Graves, Rick E. and Brian M. Argrow. 1999. Bulk viscosity: past to present. *Journal of Thermophysics and Heat Transfer* 13(3): 337–42. https://doi.org/10.2514/2.6443.

Gupta, Abhay, Amruth Bhargav and Arumugam Manthiram. 2021. Evoking high-donor-number-assisted and organosulfur-mediated conversion in lithium–sulfur batteries. *ACS Energy Letters* 6(1): 224–31. https://doi.org/10.1021/acsenergylett.0c02461.

Gupta, Jyoti K., Dave J. Adams and Neil G. Berry. 2016. Will It Gel? Successful computational prediction of peptide gelators using physicochemical properties and molecular fingerprints. *Chemical Science* 7(7): 4713–19. http://dx.doi.org/10.1039/C6SC00722H.

Gupta, Munishwar N. 1992. Enzyme function in organic solvents. *European Journal of Biochemistry* 203(1-2): 25–32. https://doi.org/10.1111/j.1432-1033.1992.tb19823.x.

Hao Caixia, Zhao Long, Yue Xiaoqing, Pang Yujie and Zhang Jianbin. 2019. Density, dynamic viscosity, excess properties and intermolecular interaction of triethylene glycol + n,n-dimethylformamide binary mixture. *Journal of Molecular Liquids* 274: 730–39. http://www.sciencedirect.com/science/article/pii/S0167732218343046.

Hao Jinkun, Cheng He, Butler Paul, Zhang Li, Han and C. Charles. 2010. Origin of cononsolvency, based on the structure of tetrahydrofuran-water mixture. *The Journal of Chemical Physics* 132(15): 154902. https://doi.org/10.1063/1.3381177.

Hasell Tom, Culshaw Jamie L., Chong Samantha Y., Schmidtmann Marc, Little Marc A., Jelfs Kim E. et al. 2014. Controlling the crystallization of porous organic cages: molecular analogs of isoreticular frameworks using shape-specific directing solvents. *Journal of the American Chemical Society* 136(4): 1438–48. https://doi.org/10.1021/ja409594s.

Heidaryan, Ehsan, Tahmas Hatami, Masoud Rahimi and Jamshid Moghadasi. 2011. Viscosity of pure carbon dioxide at supercritical region: measurement and correlation approach. *The Journal of Supercritical Fluids* 56(2): 144–51. http://www.sciencedirect.com/science/article/pii/S0896844610005127.

Heintza, Juliette, Irea Touche, Moises Teles dos Santos and Vincent Gerbaud. 2012. An integrated framework for product formulation by computer aided mixture design. pp. 702–6. *In*: Ian David Lockhart Bogle and B.T. Michael (eds.). *22 European Symposium on Computer Aided Process Engineering* - Computer Aided Chemical Engineering Fairweather. Elsevier. http://www.sciencedirect.com/science/article/pii/B9780444595195501416.

Held Christoph, Reschke Thomas, Mohammad Sultan, Luza Armando and Sadowski Gabriele. 2014. EPC-SAFT revised. *Chemical Engineering Research and Design* 92(12): 2884–97. https://www.sciencedirect.com/science/article/pii/S0263876214002469.

Heldebrant David J., Witt Heather N., Walsh Sarah M., Ellis Taryn, Rauscher Japheth and Jessop Philip G. 2006. Liquid polymers as solvents for catalytic reductions. *Green Chemistry* 8(9): 807–15. http://dx.doi.org/10.1039/B605405F.

Heyda, Jan, Anja Muzdalo and Joachim Dzubiella. 2013. Rationalizing polymer swelling and collapse under attractive cosolvent conditions. *Macromolecules* 46(3): 1231–38. https://doi.org/10.1021/ma302320y.

Hintermair, Ulrich, Walter Leitner and Philip Jessop. 2010. Expanded liquid phases in catalysis: gas-expanded liquids and liquid–supercritical fluid biphasic systems. *Handbook of Green Chemistry*. https://doi.org/10.1002/9783527628698.hgc037.

Hirst, Andrew R. and David K. Smith. 2004. Solvent effects on supramolecular gel-phase materials: two-component dendritic gel. *Langmuir* 20(25): 10851–57. https://doi.org/10.1021/la048178c.

Hoang Hai Nam, Granero-Fernandez Emanuel, Shinjiro Yamada, Shuichi Mori, Kagechika Hiroyuki and Medina-Gonzalez Yaocihuatl. 2017. Modulating biocatalytic activity toward sterically bulky substrates in CO<inf>2</Inf>-expanded biobased liquids by tuning the physicochemical properties. *ACS Sustainable Chemistry and Engineering* 5(11): 11051–11059. http://dx.doi.org/10.1021/acssuschemeng.7b03018.

Hopp-Hirschler, Manuel and Ulrich Nieken. 2018. Modeling of pore formation in phase inversion processes: model and numerical results. *Journal of Membrane Science* 564: 820–31. http://www.sciencedirect.com/science/article/pii/S0376738818310676.

Huang Jianfei, Luong Hoang Mai, Lee Jaewon, Chae Sangmin, Yi Ahra, Qu Zhong-Ze et al. 2023. Green-solvent-processed high-performance broadband organic photodetectors. *ACS Applied Materials & Interfaces* 15(31): 37748–55. https://doi.org/10.1021/acsami.3c09391.

Huber, M.L., E.A. Sykioti, M.J. Assael and R.A. Perkins. 2016. Reference correlation of the thermal conductivity of carbon dioxide from the triple point to 1100 K and up to 200 MPa. *Journal of Physical and Chemical Reference Data* 45(1): 13102. https://www.ncbi.nlm.nih.gov/pubmed/27064300.

Hyman, Anthony A., Christoph A. Weber and Frank Jülicher. 2014. Liquid-liquid phase separation in biology. *Annual Review of Cell and Developmental Biology* 30(1): 39–58. https://doi.org/10.1146/annurev-cellbio-100913-013325.

Hyun Park, Sang, In Su Jin and Jae Woong Jung. 2021. Green solvent engineering for environment-friendly fabrication of high-performance perovskite solar cells. *Chemical Engineering Journal* 425: 131475. https://www.sciencedirect.com/science/article/pii/S1385894721030564.

Ichikawa, Takahiro, Takashi Kato and Hiroyuki Ohno. 2019. Dimension control of ionic liquids. *Chemical Communications* 55(57): 8205–14. http://dx.doi.org/10.1039/C9CC04280F.

Imre, A.R., G. Hazi, A. Horvath, Cs. Maraczy, V. Mazur and S. Artemenko. 2011. The effect of low-concentration inorganic materials on the behaviour of supercritical water. *Nuclear Engineering and Design* 241(1): 296–300. http://www.sciencedirect.com/science/article/pii/S0029549310007569.

Imre, A.R., C. Ramboz, U.K. Deiters and T. Kraska. 2015. Anomalous fluid properties of carbon dioxide in the supercritical region: application to geological CO_2 storage and related hazards. *Environmental Earth Sciences* 73(8): 4373–84. https://doi.org/10.1007/s12665-014-3716-5.

Ishmael Mitchell P.E., B. Stutzman Mauren, Z. Lukawski Maciej, A. Escobedo Fernando and W. Tester Jefferson. 2017. Heat capacities of supercritical fluid mixtures: comparing experimental measurements with monte carlo molecular simulations for carbon dioxide-methanol mixtures. *The Journal of Supercritical Fluids* 123: 40–49. http://www.sciencedirect.com/science/article/pii/S0896844616304545.

Jaeger, Frederike, Omar K. Matar and Erich A. Müller. 2018. Bulk viscosity of molecular fluids. *The Journal of Chemical Physics* 148(17): 174504. https://doi.org/10.1063/1.5022752.

James, Stuart L. 2016. The dam bursts for porous liquids. *Advanced Materials* 28(27): 5712–16. https://doi.org/10.1002/adma.201505607.

Jeon Nam Joong, Noh Jun hong, Kim Young chan, Yang Woon Seok, Ryu Seungchan and Seok Sang II. 2014. Solvent engineering for high-performance inorganic–organic hybrid perovskite solar cells. *Nature Materials* 13: 897. https://doi.org/10.1038/nmat4014.

Jessop Philip G., J. Heldebrant David, Li Xiaowang, A. Eckert Charles and L. Liotta Charles. 2005. Reversible nonpolar-to-polar solvent. *Nature* 436(7054): 1102. https://doi.org/10.1038/4361102a.

Jessop, Philip G., Sean M. Mercer and David J. Heldebrant. 2012. CO_2-triggered switchable solvents, surfactants, and other materials. *Energy & Environmental Science* 5(6): 7240–53. http://dx.doi.org/10.1039/C2EE02912J.

Jessop, Philip G. and Bala Subramaniam. 2007. Gas-expanded liquids. *Chemical Reviews* 107(6): 2666–94. http://dx.doi.org/10.1021/cr040199o.

Jia Di, Zuo Taisen, Rogers Sarah, Cheng He, Hammouda Boualem and Han Charles C. 2016. Re-entrance of poly(N,N-diethylacrylamide) in D_2O/d-Ethanol mixture at 27°C. *Macromolecules* 49(14): 5152–59. https://doi.org/10.1021/acs.macromol.6b00785.

Jia Hao, Kim Ju-Myung, Gao Peiyuan, Xu Yaobin, Engelhard Mark H., Matthews Bethany E. et al. 2023. A systematic study on the effects of solvating solvents and additives in localized high-concentration electrolytes over electrochemical performance of lithium-ion batteries. *Angewandte Chemie International Edition* 62(17): e202218005. https://doi.org/10.1002/anie.202218005.

Jiao Jieming, Yang Chenguang, Wang Zhen, Yan Chang and Fang Changqing. 2023. Solvent engineering for the formation of high-quality perovskite films: a review. *Results in Engineering* 18: 101158. https://www.sciencedirect.com/science/article/pii/S2590123023002852.

Jirasek, Fabian and Hans Hasse. 2023. Combining machine learning with physical knowledge in thermodynamic modeling of fluid mixtures. *Annual Review of Chemical and Biomolecular Engineering* 14(1): 31–51. https://doi.org/10.1146/annurev-chembioeng-092220-025342.

Juilland, P. and E. Gallucci. 2015. Morpho-topological investigation of the mechanisms and kinetic regimes of alite dissolution. *Cement and Concrete Research* 76: 180–91. https://www.sciencedirect.com/science/article/pii/S0008884615001702.

Juilland, Patrick, Emmanuel Gallucci, Robert Flatt and Karen Scrivener. 2010. Dissolution theory applied to the induction period in alite hydration. *Cement and Concrete Research* 40(6): 831–44. https://www.sciencedirect.com/science/article/pii/S000888461000027X.

Julianto, T.S. and Suratmi. 1823. The effect of concentrations and volumes of methanol in reducing free fatty acid content of used cooking oil as biodiesel feedstock. International Conference on Chemistry.

Jung, Jennifer and Michel Perrut. 2001. Particle design using supercritical fluids: literature and patent survey. *The Journal of Supercritical Fluids* 20(3): 179–219. http://www.sciencedirect.com/science/article/pii/S089684460100064X.

Kafle, Arjun, Jun Liu and Lina Cui. 2016. Controlling the stereoselectivity of glycosylation via solvent effects. *Canadian Journal of Chemistry* 94(11): 894–901. https://doi.org/10.1139/cjc-2016-0417.

Kalinichev, A.G. and S.V. Churakov. 1999. Size and topology of molecular clusters in supercritical water: a molecular dynamics simulation. *Chemical Physics Letters* 302(5): 411–17. http://www.sciencedirect.com/science/article/pii/S0009261499001748.

Kamlet, Mortimer J., Jose Luis M. Abboud, Michael H. Abraham and R.W. Taft. 1983. Linear solvation energy relationships. 23. A comprehensive collection of the solvatochromic parameters, Π^*, α, and β, and some methods for simplifying the generalized solvatochromic equation. *The Journal of Organic Chemistry* 48(17): 2877–87. https://doi.org/10.1021/jo00165a018.

Kamlet, Mortimer J., Jose Luis Abboud and R.W. Taft. 1977. The solvatochromic comparison method. 6. The Π^* scale of solvent polarities. *Journal of the American Chemical Society* 99(18): 6027–38. https://doi.org/10.1021/ja00460a031.

Kamlet, Mortimer J., Thomas N. Hall, John Boykin and R.W. Taft. 1979a. Linear solvation energy relationships. 6. Additions to and correlations with the Pi.* Scale of solvent polarities. *The Journal of Organic Chemistry* 44(15): 2599–2604. https://doi.org/10.1021/jo01329a001.

Kamlet, Mortimer J., Mary Elizabeth Jones, Robert W. Taft and José-Luis Abboud. 1979b. Linear solvation energy relationships. Part 2. Correlations of electronic spectral data for aniline indicators with solvent Π* and β values. *Journal of the Chemical Society, Perkin Transactions* 2(3): 342–48. http://dx.doi.org/10.1039/P29790000342.

Kamlet, Mortimer J. and R.W. Taft. 1976. The solvatochromic comparison method. I. The Beta.-Scale of solvent hydrogen-bond acceptor (HBA) basicities. *Journal of the American Chemical Society* 98(2): 377–83. https://doi.org/10.1021/ja00418a009.

Kanagaraj, Kuppusamy, M. Alagesan, Yoshihisa Inoue and C. Yang. 2016. Solvation effects in supramolecular chemistry. *In: Reference Module in Chemistry, Molecular Sciences and Chemical Engineering.*

Kaner, Papatya, Alexander V. Dudchenko, Meagan S. Mauter and Ayse Asatekin. 2019. Zwitterionic copolymer additive architecture affects membrane performance: fouling resistance and surface rearrangement in saline solutions. *Journal of Materials Chemistry A* 7(9): 4829–46. http://dx.doi.org/10.1039/C8TA11553B.

Katritzky Alan R., Tamm Tarmo, Wang Yilin, Sild Sulev and Karelson Mati. 1999. QSPR treatment of solvent scales. *Journal of Chemical Information and Computer Sciences* 39(4): 684–91. https://doi.org/10.1021/ci980225h.

Katritzky Alan R., Fara Dan C., Yang Hongfang, Tämm Kaido, Tamm Tamo and Karelson Mati. 2004. Quantitative measures of solvent polarity. *Chem. Rev.* 104: 175–198.

Katritzky Alan R., Fara Dan C., Kuanar Minati, Hur Evrim and Karelson Mati. 2005. The classification of solvents by combining classical QSPR methodology with principal component analysis. *The Journal of Physical Chemistry A* 109(45): 10323–10341.

Ke Weijun, Mao Lingling, Stoumpos Constantinos C., Hoffman Justin, Spanopoulos Ioannis, Mohite Aditya D. et al. 2019. Compositional and solvent engineering in dion–jacobson 2D perovskites boosts solar cell efficiency and stability. *Advanced Energy Materials* 9(10): 1803384. https://doi.org/10.1002/aenm.201803384.

Kearsey Rachel, J., M. Alston Ben, E. Briggs Michael, L. Greenaway Rebecca and I. Cooper Andrew. 2019. Accelerated robotic discovery of Type II porous liquids. *Chemical Science* 10(41): 9454–65. http://dx.doi.org/10.1039/C9SC03316E.

Khaw, Kooi-Yeong, Marie-Odile Parat, Paul Nicholas Shaw and James Robert Falconer. 2017. Solvent supercritical fluid technologies to extract bioactive compounds from natural sources: a review. *Molecules (Basel, Switzerland)* 22(7): 1186. https://pubmed.ncbi.nlm.nih.gov/28708073.

Khmelnitsky, Yuri L., Stephanie H. Welch, Douglas S. Clark and Jonathan S. Dordick. 1994. Salts dramatically enhance activity of enzymes suspended in organic solvents. *Journal of the American Chemical Society* 116(6): 2647–48. https://doi.org/10.1021/ja00085a066.

Khokarale, Santosh Govind and Jyri-Pekka Mikkola. 2019. Efficient and catalyst free synthesis of acrylic plastic precursors: methyl propionate and methyl methacrylate synthesis through reversible CO_2 capture. *Green Chemistry* 21(8): 2138–47. http://dx.doi.org/10.1039/C9GC00413K.

Kim, Jeong-Hoon and Kew-Ho Lee. 1998. Effect of PEG additive on membrane formation by phase inversion. *Journal of Membrane Science* 138(2): 153–63. http://www.sciencedirect.com/science/article/pii/S037673889700224X.

Klibanov Alexander M. 2001. Improving enzymes by using them in organic solvents. *Nature* 11(409): 241–246.

Knebel Alexander, Bavykina Anastasia, Datta Shuvo Jit, Sundermann Lion, Garzon-Tovar Luis, Lebedev Yury et al. 2020. Solution processable metal–organic frameworks for mixed matrix membranes using porous liquids. *Nature Materials* 19(12): 1346–53. https://doi.org/10.1038/s41563-020-0764-y.

Knez, Željko, Darija Cör and Maša Knez Hrnčič. 2018. Solubility of solids in sub- and supercritical fluids: a review 2010–2017. *Journal of Chemical & Engineering Data* 63(4): 860–84. https://doi.org/10.1021/acs.jced.7b00778.

Knez, Željko, Maja Leitgeb and Mateja Primožič. 2018. Chemical reactions in subcritical and supercritical fluids BT - Encyclopedia of Sustainability Science and Technology. pp. 1–21. *In*: Robert A. Meyers (ed.). New York, NY: Springer New York. https://doi.org/10.1007/978-1-4939-2493-6_1004-1.

Kojima, Hiroyuki, Fumihiko Tanaka, Christine Scherzinger and Walter Richtering. 2013. Temperature dependent phase behavior of PNIPAM microgels in mixed water/methanol solvents. *Journal of Polymer Science Part B: Polymer Physics* 51(14): 1100–1111. https://doi.org/10.1002/polb.23194.

Komal, Gagandeep Singh, Gurbir Singh and Tejwant Singh Kang. 2018. Aggregation behavior of sodium dioctyl sulfosuccinate in deep eutectic solvents and their mixtures with water: an account of solvent's polarity, cohesiveness, and solvent structure. *ACS Omega* 3(10): 13387–98. https://doi.org/10.1021/acsomega.8b01637.

Korevaar Peter A., J. George Subi, J. Markvoort Albert, M.J. Smulders Maarten, A.J. Hilbers Peter, P.H. Schenning Albert et al. 2012. Pathway complexity in supramolecular polymerization. *Nature* 481(7382): 492–96. https://doi.org/10.1038/nature10720.

Korevaar, Peter A., Charley Schaefer, Tom F.A. de Greef and E.W. Meijer. 2012. Controlling chemical self-assembly by solvent-dependent dynamics. *Journal of the American Chemical Society* 134(32): 13482–91. https://doi.org/10.1021/ja305512g.

Korlepara, Divya B. and S. Balasubramanian. 2018. Molecular modelling of supramolecular one dimensional polymers. *RSC Advances* 8(40): 22659–69. http://dx.doi.org/10.1039/C8RA03402H.

Kosower, Edward M. 1958. The effect of solvent on spectra. I. A new empirical measure of solvent polarity: Z-values. *Journal of the American Chemical Society* 80(13): 3253–60. https://doi.org/10.1021/ja01546a020.

Kosower, Edward M. and Brian G. Ramsey. 1959. The effect of solvent on spectra. IV. Pyridinium cyclopentadienylide. *Journal of the American Chemical Society* 81(4): 856–60. https://doi.org/10.1021/ja01513a028.

Kraft, Stephan and Frédéric Vogel. 2017. Estimation of binary diffusion coefficients in supercritical water: mini review. *Industrial & Engineering Chemistry Research* 56(16): 4847–55. https://doi.org/10.1021/acs.iecr.7b00382.

Kumar, Ashok, Kartik Dhar, Shamsher Singh Kanwar and Pankaj Kumar Arora. 2016. Lipase catalysis in organic solvents: advantages and applications. *Biological Procedures Online* 18(1): 2. https://doi.org/10.1186/s12575-016-0033-2.

Kwon Nayoon, Lee Jaehee, Ko Min Jae, Kim Young Yun and Seao Jangwon. 2023. Recent progress of eco-friendly manufacturing process of efficient perovskite solar cells. *Nano Convergence* 10(1): 28. https://doi.org/10.1186/s40580-023-00375-5.

Lai Beibei, Cahir John, Tsang Min Ying, Jacquemin Johan, Rooney David, Murrer Barry et al. 2021. Type 3 porous liquids for the separation of ethane and ethene. *ACS Applied Materials & Interfaces* 13(1): 932–36. https://doi.org/10.1021/acsami.0c19044.

Lakshmi, T.S. and P.K. Nandi. 1976. Effects of sugar solutions on the activity coefficients of aromatic amino acids and their N-acetyl ethyl esters. *The Journal of Physical Chemistry* 80(3): 249–52. https://doi.org/10.1021/j100544a008.

Lan, Y., M.G. Corradini, R.G. Weiss, S.R. Raghavan and M.A. Rogers. 2015. To gel or not to gel: correlating molecular gelation with solvent parameters. *Chemical Society Reviews* 44(17): 6035–58. http://dx.doi.org/10.1039/C5CS00136F.

Langenfeld, John J., Steven B. Hawthorne, David J. Miller and Joseph. Tehrani. 1992. Method for determining the density of pure and modified supercritical fluids. *Analytical Chemistry* 64(19): 2263–66. https://doi.org/10.1021/ac00043a014.

Lasaga, Antonio C. and Alex E. Blum. 1986. Surface chemistry, etch pits and mineral-water reactions. *Geochimica et Cosmochimica Acta* 50(10): 2363–79. https://www.sciencedirect.com/science/article/pii/0016703786900888.

Lawrenson, Stefan B. 2018. Greener solvents for solid-phase organic synthesis. 90(1): 157–65. https://doi.org/10.1515/pac-2017-0505.

Lazzaroni Michael, J., Bush David, A. Eckert Charles, C. Franck Timothy, Gupta Sumnesh and D. Olson James. 2005. Revision of MOSCED parameters and extension to solid solubility calculations. *Industrial & Engineering Chemistry Research* 44(11): 4075–83. https://doi.org/10.1021/ie049122g.

Leron, Rhoda B., Allan N. Soriano and Meng-Hui Li. 2012. Densities and refractive indices of the deep eutectic solvents (choline chloride+ethylene glycol or glycerol) and their aqueous mixtures at the temperature ranging from 298.15 to 333.15 K. *Journal of the Taiwan Institute of Chemical Engineers* 43(4): 551–57. https://www.sciencedirect.com/science/article/pii/S1876107012000089.

Li, Cheng-Peng and Miao Du. 2011. Role of solvents in coordination supramolecular systems. *Chemical Communications* 47(21): 5958–72. http://dx.doi.org/10.1039/C1CC10935A.

Li, Qinyi and Zhang Xing. 2014. Raman spectra method for determining viscosity of supercritical fluids. pp. 8837–50. *In*: *Proceedings of the 15th International Heat Transfer Conference*.

Licence Peter, Ke Jie, Sokolova Maia, K. Ross Stephen and Poliakoff Martyn. 2003. Chemical reactions in supercritical carbon dioxide: from laboratory to commercial plant. *Green Chemistry* 5(2): 99–104. http://dx.doi.org/10.1039/B212220K.

Lilichenko, Mark and Dmitry V. Matyushov. 2003. Control of electron transfer rates in liquid crystalline media. *The Journal of Physical Chemistry B* 107(9): 1937–40. https://doi.org/10.1021/jp026688e.

Lim, Hyuntae and YounJoon Jung. 2019. Delfos: Deep learning model for prediction of solvation free energies in generic organic solvents. *Chemical Science* 10(36): 8306–15. http://dx.doi.org/10.1039/C9SC02452B.

Lin, Shiang-Tai, Chieh-Ming Hsieh and Ming-Tsung Lee. 2007. Solvation and chemical engineering thermodynamics. *Journal of the Chinese Institute of Chemical Engineers* 38(5): 467–76. https://www.sciencedirect.com/science/article/pii/S0368165307000883.

Liu Feijie, Xue Zhimin, Lan Xue, Liu Zhenghui and Mu Tiencheng. 2021. CO_2 switchable deep eutectic solvents for reversible emulsion phase separation. *Chemical Communications* 57(5): 627–30. http://dx.doi.org/10.1039/D0CC06963A.

Liu Xinghua, Yan Kang, Tan Dawei, Liang Xiao, Zhang Hongmei and Huang Wei. 2018. Solvent engineering improves efficiency of lead-free tin-based hybrid perovskite solar cells beyond 9%. *ACS Energy Letters* 3(11): 2701–7. https://doi.org/10.1021/acsenergylett.8b01588.

Van Lomlmel Ruben, Zhao Jianyu, De Boirggraeve Wim M., De Proft Frank and Alonso Mercedes. 2020. Molecular dynamics based descriptors for predicting supramolecular gelation. *Chemical Science* 11(16): 4226–38. http://dx.doi.org/10.1039/D0SC00129E.

Lorimer, J.W. 1993. Thermodynamics of solubility in mixed solvent systems. 65(2): 183–91. https://doi.org/10.1351/pac199365020183.

Lou, Xianwen, Hans-Gerd Janssen and Carel A. Cramers. 1997. Temperature and pressure effects on solubility in supercritical carbon dioxide and retention in supercritical fluid chromatography. *Journal of Chromatography A* 785(1): 57–64. http://www.sciencedirect.com/science/article/pii/S0021967397006936.

Luo, C.J., E. Stride and M. Edirisinghe. 2012. Mapping the influence of solubility and dielectric constant on electrospinning polycaprolactone solutions. *Macromolecules* 45(11): 4669–80. https://doi.org/10.1021/ma300656u.

Luo, Yi, Chengzhe Yin and Leming Ou. 2023. A novel green process for recovery of transition metal ions from used lithium-ion batteries by selective coordination of ethanol. *ACS Sustainable Chemistry & Engineering* 11(32): 11834–42. https://doi.org/10.1021/acssuschemeng.3c01537.

Ma, Xingjuan, Jiaqi Kong, Wei Wang and Xin Li. 2023. Green-solvent engineering for depositing qualified phenyl-C61-butyl acid methyl ester films for inverted flexible perovskite solar cells. *ACS Applied Materials & Interfaces* 15(1): 1042–52. https://doi.org/10.1021/acsami.2c17694.

Mabesoone, Mathijs F.J., Anja R.A. Palmans and E.W. Meijer. 2020. Solute–solvent interactions in modern physical organic chemistry: supramolecular polymers as a muse. *Journal of the American Chemical Society* 142(47): 19781–98. https://doi.org/10.1021/jacs.0c09293.

Machida Hiroshi, Ando Ryuya, Esaki Takehiro, Yamaguchi Tsuyoshi, Horizoe Hirotoshi, Kishimoto Akira et al. 2018. Low temperature swing process for CO_2 absorption-desorption using phase separation CO_2 capture solvent. *International Journal of Greenhouse Gas Control* 75: 1–7. http://www.sciencedirect.com/science/article/pii/S1750583617309969.

Maciejewska, Paola R. Campodónico and Daniel Glossman-Mitnik, Magdalena. 2020. Solvent effect on a model of S_NAr reaction in conventional and non-conventional solvents. *In*: Rijeka: IntechOpen, Ch. 12. https://doi.org/10.5772/intechopen.89838.

Magda, J.J., G.H. Fredrickson, R.G. Larson and E. Helfand. 1988. Dimensions of a polymer chain in a mixed solvent. *Macromolecules* 21(3): 726–32. https://doi.org/10.1021/ma00181a029.

Mahdavi Hamidreza, Zhang Huacheng, Macreadie Lauren, C.M. Doherty, Achrya Durga, J.D. Smith Stefan et al. 2022. Underlying solvent-based factors that influence permanent porosity in porous liquids. *Nano Research* 15(4): 3533–38. https://doi.org/10.1007/s12274-021-3862-5.

Mahdavi, Hamidreza, Stefan J.D. Smith, Xavier Mulet and Matthew R. Hill. 2022. Practical considerations in the design and use of porous liquids. *Materials Horizons* 9(6): 1577–1601. http://dx.doi.org/10.1039/D1MH01616D.

Mahdavi, Hossein, B.O. Mahdi Sadeghzadeh and Nader Taheri Qazvini. 2009. Phase behavior study of Poly(N-Tertbutylacrylamide-Co-Acrylamide) in the mixture of water–methanol: the role of polymer–nonsolvent second-order interactions. *Journal of Polymer Science Part B: Polymer Physics* 47(5): 455–62. https://doi.org/10.1002/polb.21650.

Malhotra, R. and L.A. Woolf. 1991. Thermodynamic properties of propanone (acetone) at temperatures from 278 K to 323 K and pressures up to 400 MPa. *The Journal of Chemical Thermodynamics* 23(9): 867–76. http://www.sciencedirect.com/science/article/pii/S0021961405802824.

Malik, Prakash Kumar, Madhusmita Tripathy, Aravind Babu Kajjam and Sabita Patel. 2020. Preferential solvation of p-nitroaniline in a binary mixture of chloroform and hydrogen bond acceptor solvents: the role of specific solute–solvent hydrogen bonding. *Physical Chemistry Chemical Physics* 22(6): 3545–62. http://dx.doi.org/10.1039/C9CP05772B.

Marchetti, M.C., S. Ramaswamy Joanny, T.B. Liverpool, J. Prost, Rao Madan et al. 2013. Hydrodynamics of soft active matter. *Reviews of Modern Physics* 85(3): 1143–89. https://link.aps.org/doi/10.1103/RevModPhys.85.1143.

Marcus, Y., M.J. Kamlet and R.W. Taft. 1988. Linear solvation energy relationships: standard molar gibbs free energies and enthalpies of transfer of ions from water into nonaqueous solvents. *The Journal of Physical Chemistry* 92(12): 3613–22. https://doi.org/10.1021/j100323a057.

Marcus, Yitzhak. 2002. *Solvent Mixtures: Properties and Selective Solvation.* 1st Ed. ed. CRC Press. Taylor & Francis.

Marcus, Yizhak. 2006. Are solubility parameters relevant to supercritical fluids? *The Journal of Supercritical Fluids* 38(1): 7–12. http://www.sciencedirect.com/science/article/pii/S089684460500241X.

Martin, Pablo, Juan J. Gaitero, Jorge S. Dolado and Hegoi Manzano. 2020. KIMERA: A kinetic montecarlo code for mineral dissolution. *Minerals* 10(9).

Martins, Marcos A.P., Zimmer Géorgia C., Rodrigues Leticia V., Orlando Tainara, Buriol Lilian, Alajarin Mateo et al. 2017. Competition between the donor and acceptor hydrogen bonds of the threads in the formation of [2]rotaxanes by clipping reaction. *New Journal of Chemistry* 41(22): 13303–18. http://dx.doi.org/10.1039/C7NJ02443F.

Mary T. Gilbert. 1987. *High Performance Liquid Chromatography.* Elsevier. https://doi.org/10.1016/C2013-0-11934-4.

McKelvey, Scott A. and William J. Koros. 1996. Phase separation, vitrification, and the manifestation of macrovoids in polymeric asymmetric membranes. *Journal of Membrane Science* 112(1): 29–39. http://www.sciencedirect.com/science/article/pii/0376738895001972.

Medina-Gonzalez, Y., P. Aimar, J.-F. Lahitte and J.-C. Remigy. 2011. Towards green membranes: preparation of cellulose acetate ultrafiltration membranes using methyl lactate as a biosolvent. *International Journal of Sustainable Engineering* 4(1).

Melchert, Wanessa R. and Fabio R.P. FábioRocha. 2016. Cloud point extraction in flow-based systems. *Reviews in Analytical Chemistry* 35(2): 41–52.

Melnyk, A., J. Namieśnik and L. Wolska. 2015. Theory and recent applications of coacervate-based extraction techniques. *TrAC Trends in Analytical Chemistry* 71: 282–92. http://www.sciencedirect.com/science/article/pii/S0165993615001417.

de Melo, M.M.R., A.J.D. Silvestre and C.M. Silva. 2014. Supercritical fluid extraction of vegetable matrices: applications, trends and future perspectives of a convincing green technology. *The Journal of Supercritical Fluids* 92: 115–76. http://www.sciencedirect.com/science/article/pii/S0896844614000928.

Mendis, Nethrue Pramuditha, Jiayuan Wang and Richard Lakerveld. 2020. A thermodynamic approach for simultaneous solvent and process design of continuous reactive crystallization with recycling. pp. 805–10. *In*: Sauro Pierucci, Flavio Manenti, Giulia Luisa Bozzano, and Davide B.T. (eds.). *30 European Symposium on Computer Aided Process Engineering* - Computer Aided Chemical Engineering Manca. Elsevier. https://www.sciencedirect.com/science/article/pii/B978012823377150135X.

Mendis, Nethrue Pramuditha, Jiayuan Wang and Richard Lakerveld. 2022. Simultaneous solvent selection and process design for continuous reaction–extraction–crystallization systems. *Industrial & Engineering Chemistry Research* 61(31): 11504–17. https://doi.org/10.1021/acs.iecr.1c05012.

Mercer, Sean M. and Philip G. Jessop. 2010. 'Switchable Water': Aqueous solutions of switchable ionic strength. *ChemSusChem* 3(4): 467–70. https://doi.org/10.1002/cssc.201000001.

Miao Yanfeng, Ren Meng, Chen Yuetian, Wang Haifei, Chen Haoran, Liu Xiaomin et al. 2023. Green solvent enabled scalable processing of perovskite solar cells with high efficiency. *Nature Sustainability* 6(11): 1465–73. https://doi.org/10.1038/s41893-023-01196-4.

Middaugh, Charles, W. Tisel, R. Haire and A. Rosenberg. 1979. Determination of the apparent thermodynamic activities of saturated protein solution. *The Journal of Biological Chemistry* 254: 367–70.

Mishra, Snigdha, Timothy Hunter, Kamal Kishore Pant and David Harbottle. 2024. Green deep eutectic solvents (DESs) for sustainable metal recovery from thermally treated PCBs: A greener alternative to conventional methods. *ChemSusChem* n/a(n/a): e202301418. https://doi.org/10.1002/cssc.202301418.

Miyata, Takashi, Hiroshi Yamada and Tadashi Uragami. 2001. Surface modification of microphase-separated membranes by fluorine-containing polymer additive and removal of dilute benzene in water through these membranes. *Macromolecules* 34(23): 8026–33. https://doi.org/10.1021/ma0107071.

Mohammad, Ali and Inamuddin (eds.). 2012. *Green Solvents I: Properties and Applications in Chemistry*. Springer Berlin Heidelberg.

Moity Laurianne, Molinier Valérie, Benazzouz Adrien, Barone René, Marion Philippe and Aubry Jean-Mary. 2014. *In silico* design of bio-based commodity chemicals: application to itaconic acid based solvents. *Green Chemistry* 16(1): 146–60. http://dx.doi.org/10.1039/C3GC41442F.

Moity, L., V. Molinier, A. Benazzouz, B. Joossen, V. Gerbaud and J-M. Aubry. 2016. A 'Top-down' *in silico* approach for designing ad hoc bio-based solvents: application to glycerol-derived solvents of nitrocellulose. *Green Chemistry* 18(11): 3239–49. http://dx.doi.org/10.1039/C6GC00112B.

Mondal, Balaka and Govardhan Reddy. 2019. Cosolvent effects on the growth of protein aggregates formed by a single domain globular protein and an intrinsically disordered protein. *The Journal of Physical Chemistry B* 123(9): 1950–60. https://doi.org/10.1021/acs.jpcb.8b11128.

Moreira Luís, Reis Marina, Elvas-Leitao Ruben, H. Abraham Michael and Martins Filomena. 2019. Quantifying solvent effects through QSPR: A new look over different model equations. *Journal of Molecular Liquids* 291: 111244. http://www.sciencedirect.com/science/article/pii/S0167732219329435.

Morgenstern, D.A., R.M. LeLacheur, D.K. Morita, M.F. Gross, M.J. Burk and W. Tumas. 1996. Green Chemistry: Designing Chemistry for the Environment. pp. 132–51. *In*: Anastas, T.C. and P.T. Williamson. Washington D. C.: American Chemical Society.

Morisue, Mitsuhiko and Ikuya Ueno. 2018. Preferential solvation unveiled by anomalous conformational equilibration of porphyrin dimers: nucleation growth of solvent–solvent segregation. *The Journal of Physical Chemistry B* 122(20): 5251–59. https://doi.org/10.1021/acs.jpcb.8b02558.

Mostert, R., H.R. van den Berg, P.S. van der Gulik and J.V. Sengers. 1990. The thermal conductivity of ethane in the critical region. *The Journal of Chemical Physics* 92(9): 5454–62. https://doi.org/10.1063/1.458523.

Mostofian Barmak, M. Cai Charles, Smith Micholas Dean, Petridis Loukas, Cheng Xialoin, E. Wyman Charles et al. 2016. Local phase separation of co-solvents enhances pretreatment of biomass for bioenergy applications. *Journal of the American Chemical Society* 138(34): 10869–78. https://doi.org/10.1021/jacs.6b03285.

Mukai Masaru, Minamikawa Hiroyuki, Aoyagi Masaru, Asakawa Masumi, Shimizu Toshimi and Kogiso Masaki. 2012. Solvent-chirality selective organogelation by chiral aspartame lipids. *Soft Matter* 8(48): 11979–81. http://dx.doi.org/10.1039/C2SM26991K.

Mukherji Debashish, Wagner Manfred, D. Watson Mark, Winzen Svenja, E. de Oliveira tiago, M. Marques Carlos et al. 2016. Relating side chain organization of PNIPAm with its conformation in aqueous methanol. *Soft Matter* 12(38): 7995–8003. http://dx.doi.org/10.1039/C6SM01789D.

Mukherji Debashish, Watson Mark D., Morbasch Svenja, Schmutz Marc, Wagner Manfred, Marques Carlos M. et al. 2019. Soft and smart: co-nonsolvency-based design of multiresponsive copolymers. *Macromolecules* 52(9): 3471–78. https://doi.org/10.1021/acs.macromol.9b00414.

Mukherji, Debashish, Carlos M. Marques and Kurt Kremer. 2014. Polymer collapse in miscible good solvents is a generic phenomenon driven by preferential adsorption. *Nature Communications* 5(1): 4882. https://doi.org/10.1038/ncomms5882.

Mukherji, Debashish, Carlos M. Marques and Kurt Kremer, 2017. Collapse in two good solvents, swelling in two poor solvents: defying the laws of polymer solubility? *Journal of Physics: Condensed Matter* 30(2): 24002. http://dx.doi.org/10.1088/1361-648X/aa9c77.

Mulvihill, Martin J., Evan S. Beach, Julie B. Zimmerman and Paul T. Anastas. 2011. Green chemistry and green engineering: a framework for sustainable technology development. *Annual Review of Environment and Resources* 36(1): 271–93. https://doi.org/10.1146/annurev-environ-032009-095500.

Mushenheim, Peter C., S. Pendery Joel and B. Weibel Douglas. 2016. Straining soft colloids in aqueous nematic liquid crystals. *Proceedings of the National Academy of Sciences* 113(20): 5564–69. https://doi.org/10.1073/pnas.1600836113.

Musie, Ghezai, Ming Wei, Bala Subramaniam and Daryle H. Busch. 2001. Catalytic oxidations in carbon dioxide-based reaction media, including novel CO_2-expanded phases. *Coordination Chemistry Reviews* 219–221(0): 789–820. http://www.sciencedirect.com/science/article/pii/S0010854501003678.

Naderi, Noushin, Seyed Saeid Hosseini and Yomen Atassi. 2022. Tailoring the morphology and performance of polyacrylonitrile ultrafiltration membranes for produced water treatment via solvent mixture strategy. *Chemical Engineering & Technology* 45(9): 1655–65. https://doi.org/10.1002/ceat.202100638.

Nag, Abhijit, Carole A. Morrison and Jason B. Love. 2022. Rapid dissolution of noble metals in organic solvents. *ChemSusChem* 15(20): e202201285. https://doi.org/10.1002/cssc.202201285.

Le Neindre, B., Y. Garrabos and M. Nikravech. 2014. Measurements of the thermal conductivity of propane in the supercritical region. *Journal of Chemical & Engineering Data* 59(11): 3422–33. https://doi.org/10.1021/je500395f.

Le Neindre, B., Y. Garrabos and R. Tufeu. 1984. Thermal conductivity in supercritical fluids. *Berichte der Bunsengesellschaft für physikalische Chemie* 88(9): 916–20. https://doi.org/10.1002/bbpc.19840880930.

Nishikawa, Keiko, Asako Ayusawa Arai and Takeshi Morita. 2004. Density fluctuation of supercritical fluids obtained from small-angle x-ray scattering experiment and thermodynamic calculation. *The Journal of Supercritical Fluids* 30(3): 249–57. http://www.sciencedirect.com/science/article/pii/S0896844603002110.

Noel, Nakita K., Bernard Wenger, Severin N. Habisreutinger and Henry J. Snaith. 2022. Utilizing nonpolar organic solvents for the deposition of metal-halide perovskite films and the realization of organic semiconductor/perovskite composite photovoltaics. *ACS Energy Letters* 7(4): 1246–54. https://doi.org/10.1021/acsenergylett.2c00120.

Noroozi, Javad. 2021. Design of Improved Carbon Capture Solvents Using Molecular Simulation. UWSpace. http://hdl.handle.net/10012/17152.

Nothdurft Katja, H. Müller David, Brands Thorsten, Bardow André and Richtering Walter. 2019. Enrichment of methanol inside PNIPAM gels in the cononsolvency-induced collapse. *Physical Chemistry Chemical Physics* 21(41): 22811–18. http://dx.doi.org/10.1039/C9CP04383G.

Novak, Nefeli, Georgios M. Kontogeorgis, Marcelo Castier and Ioannis G. Economou. 2023. Mixed solvent electrolyte solutions: a review and calculations with the esaft-vr mie equation of state. *Industrial & Engineering Chemistry Research* 62(34): 13646–65. https://doi.org/10.1021/acs.iecr.3c00717.

O'Reilly, Niamh, Nicola Giri and Stuart L. James. 2007. Porous liquids. *Chemistry – A European Journal* 13(11): 3020–25. https://doi.org/10.1002/chem.200700090.

Odele, O. and S. Macchietto. 1993. Computer aided molecular design: a novel method for optimal solvent selection. *Fluid Phase Equilibria* 82: 47–54. https://www.sciencedirect.com/science/article/pii/037838129387127M.

Ogino, Hiroyasu and Haruo Ishikawa. 2001. Enzymes which are stable in the presence of organic solvents. *Journal of Bioscience and Bioengineering* 91(2): 109–16. http://www.sciencedirect.com/science/article/pii/S1389172301800517.

Ojeda, C. Bosch and F. Sánchez Rojas. 2009. Separation and preconcentration by a cloud point extraction procedure for determination of metals: an overview. *Analytical and Bioanalytical Chemistry* 394(3): 759–82. https://doi.org/10.1007/s00216-009-2660-9.

Oldham, C.E., D.E. Farrow and S. Peiffer. 2013. A generalized damköhler number for classifying material processing in hydrological systems. *Hydrol. Earth Syst. Sci.* 17(3): 1133–48. https://www.hydrol-earth-syst-sci.net/17/1133/2013/.

Oliveira Cleverson, J.F., K.S. Freitas Sunny, P.P. de Sousa Guida Igor, M. Esteves Pierre and A. Simao Renata. 2020. Solvent role on covalent organic framework thin film formation promoted by ultrasound. *Colloids and Surfaces A: Physicochemical and Engineering Aspects* 585: 124086. https://www.sciencedirect.com/science/article/pii/S0927775719310787.

Onozawa-Komatsuzaki Nobuko, Tsuchiya Daisuke, Inoue Shinichi, Kogo Atsushi, Ueno Toshiya and Takurou N. Murakami. 2023. Green-solvent-soluble, highly efficient dopant-free hole-transporting material for perovskite solar cells. *Applied Physics Express* 16(1): 16502. https://dx.doi.org/10.35848/1882-0786/acaa88.

Orakdogen, Nermin and Oguz Okay. 2006. Reentrant conformation transition in poly(N,N-dimethylacrylamide) hydrogels in water–organic solvent mixtures. *Polymer* 47(2): 561–68. http://www.sciencedirect.com/science/article/pii/S0032386105017179.

Orozco, Modesto and F. Javier Luque. 2000. Theoretical methods for the description of the solvent effect in biomolecular systems. *Chemical Reviews* 100(11): 4187–4226. https://doi.org/10.1021/cr990052a.

Pabsch, Daniel, Paul Figiel, Gabriele Sadowski and Christoph Held. 2022. Solubility of electrolytes in organic solvents: solvent-specific effects and ion-specific effects. *Journal of Chemical & Engineering Data* 67(9): 2706–18. https://doi.org/10.1021/acs.jced.2c00203.

Pagonis, Konstantinos and Georgios Bokias. 2004. Upper critical solution temperature—type cononsolvency of poly(N,N-dimethylacrylamide) in water—organic solvent mixtures. *Polymer* 45(7): 2149–53. http://www.sciencedirect.com/science/article/pii/S0032386104001193.

Pal Mahi, Rai Rewa, Yadav Anita, Khanna Rajesh, A. Baker Gary and Pandey Siddharth. 2014. Self-aggregation of sodium dodecyl sulfate within (choline chloride + urea) deep eutectic solvent. *Langmuir* 30(44): 13191–98. https://doi.org/10.1021/la5035678.

Pal, Mahi, Ranjan K. Singh and Siddharth Pandey. 2015. Evidence of self-aggregation of cationic surfactants in a choline chloride+glycerol deep eutectic solvent. *ChemPhysChem* 16(12): 2538–42. https://doi.org/10.1002/cphc.201500357.

Palacci, J., S. Sacanna, S.-H. Kim, G.-R. Yi, D.J. Pine and P.M. Chaikin. 2014. Light-activated self-propelled colloids. *Philosophical Transactions of the Royal Society A: Mathematical, Physical and Engineering Sciences* 372(2029): 20130372. https://doi.org/10.1098/rsta.2013.0372.

Palafox-Hernandez, J. Pablo, Camina H. Mendis, Ward H. Thompson and Brian B. Laird. 2019. Pressure and temperature tuning of gas-expanded liquid structure and dynamics. *The Journal of Physical Chemistry B* 123(13): 2915–24. https://doi.org/10.1021/acs.jpcb.8b09826.

Palamareva, M.D. 2005. Liquid chromatography | Principles. pp. 112–18. *In*: Paul Worsfold, Alan Townshend, and Colin B.T. (eds.). - Encyclopedia of Analytical Science (Second Edition) Poole. Oxford: Elsevier. http://www.sciencedirect.com/science/article/pii/B0123693977003149.

Paleologos, Evangelos K., Dimosthenis L. Giokas and Miltiades I. Karayannis. 2005. Micelle-mediated separation and cloud-point extraction. *TrAC Trends in Analytical Chemistry* 24(5): 426–36. http://www.sciencedirect.com/science/article/pii/S0165993605000609.

Pallewela, Gayani N. and Paul E. Smith. 2015. Preferential solvation in binary and ternary mixtures. *The Journal of Physical Chemistry B* 119(51): 15706–17. https://doi.org/10.1021/acs.jpcb.5b10139.

Papadakis, Raffaello and Ioanna Deligkiozi. 2019. *Solvent Effects in Supramolecular Systems, Solvents, Ionic Liquids and Solvent Effects*. IntechOpen.

Papadopoulos Athanasios I., Badr Sara, Chremos Alexanros, Forte Esther, Zarogiannis Theodoros, Seferlis Panos et al. 2016. Computer-aided molecular design and selection of CO_2 capture solvents based on thermodynamics, reactivity and sustainability. *Molecular Systems Design & Engineering* 1(3): 313–34. http://dx.doi.org/10.1039/C6ME00049E.

Park, Changmoon, Matt J. Carlson and William A. Goddard. 2000. Solvent effects on the secondary structures of proteins. *The Journal of Physical Chemistry A* 104(11): 2498–2503. https://doi.org/10.1021/jp9911189.

Park, Chul Ho and Jonghwi Lee. 2009. Electrosprayed polymer particles: effect of the solvent properties. *Journal of Applied Polymer Science* 114(1): 430–37. https://doi.org/10.1002/app.30498.

Pedley, T.J. and J.O. Kessler. 1990. A new continuum model for suspensions of gyrotactic micro-organisms. *Journal of Fluid Mechanics* 212: 155–82. https://www.cambridge.org/core/article/new-continuum-model-for-suspensions-of-gyrotactic-microorganisms/534F80636BCC9E48084C208C7436DC25.

Peng Jian, Yamamoto Yukinori, Hawk Jeffrey A., Lara-Curzio Edgar and Shin Dongwon. 2020. Coupling physics in machine learning to predict properties of high-temperatures alloys. *npj Computational Materials* 6(1): 141. https://doi.org/10.1038/s41524-020-00407-2.

Peng, Yueying, Kei Nishikawa and Kiyoshi Kanamura. 2022. Effects of carbonate solvents and lithium salts in high-concentration electrolytes on lithium anode. *Journal of The Electrochemical Society* 169(6): 60548. https://dx.doi.org/10.1149/1945-7111/ac797a.

Phan Lam, Chiu Daniel, Heldebrant David J., Huttenhower Hillary, Johj Ejae, Pollet Pamela et al. 2008. Switchable solvents consisting of amidine/alcohol or guanidine/alcohol mixtures. *Industrial & Engineering Chemistry Research* 47(3): 539–45. https://doi.org/10.1021/ie070552r.

Phan, L., H. Brown, J. White, A. Hodgson and P.J. Jessop. 2009. Soybean oil extraction and separation using switchable or expanded solvents. *Green Chemistry* 11(1): 53–59. http://dx.doi.org/10.1039/B810423A.

Phifer, Jeremy R., Kimberly J. Solomon, Kayla L. Young and Andrew S. Paluch. 2017. Computing MOSCED parameters of nonelectrolyte solids with electronic structure methods in SMD and SM8 continuum solvents. *AIChE Journal* 63(2): 781–91. https://doi.org/10.1002/aic.15413.

Pinault, Thomas, Benjamin Isare and Laurent Bouteiller. 2006. Solvents with similar bulk properties induce distinct supramolecular architectures. *ChemPhysChem* 7(4): 816–19. https://doi.org/10.1002/cphc.200500636.

Pioro, Igor. 2011. Thermophysical properties at critical and supercritical pressures. *In*: Sarah Mokry (ed.). ED1 - Aziz Belmiloudi. Rijeka: IntechOpen, Ch. 22. https://doi.org/10.5772/13790.

Podapangi Suresh K., Jafarzadeh Farshad, Mattiello Sara, Bhavani Korukonda Tujla, Singh Akash, Beverina Luca et al. 2023. Green solvents, materials, and lead-free semiconductors for sustainable fabrication of perovskite solar cells. *RSC Advances* 13(27): 18165–206. http://dx.doi.org/10.1039/D3RA01692G.

Pollet Pamela, Davey Evan A., Urena-Benavides Esteban, Eckert A. Charles and L. Liotta Charles. 2014. Solvents for sustainable chemical processes. *Green Chemistry* 16(3): 1034–55. http://dx.doi.org/10.1039/C3GC42302F.

Pollet, Pamela, Charles A. Eckert and Charles L. Liotta. 2011. Switchable solvents. *Chemical Science* 2(4): 609–14. http://dx.doi.org/10.1039/C0SC00568A.

Prat Denis, Pardigon Olivier, Flemming Hans-Wolfram, Letestu Sylvie, Ducandas Véronique, Isnard Pascal et al. 2013. Sanofi's solvent selection guide: a step toward more sustainable processes. *Organic Process Research & Development* 17(12): 1517–25. https://doi.org/10.1021/op4002565.

Protsenko, V.S., A.A. Kityk, D.A. Shaiderov and F.I. Danilov. 2015. Effect of water content on physicochemical properties and electrochemical behavior of ionic liquids containing choline chloride, ethylene glycol and hydrated nickel chloride. *Journal of Molecular Liquids* 212: 716–22. http://www.sciencedirect.com/science/article/pii/S0167732215308382.

Prymidis, Vasileios, Harmen Sielcken and Laura Filion. 2015. Self-assembly of active attractive spheres. *Soft Matter* 11(21): 4158–66. http://dx.doi.org/10.1039/C5SM00127G.

Qian Qihui, Asinger Patrick A., Lee Moon Joo, Han, Gang, Rodriguez Mizrahi Katherine, Lin Sharon et al. 2020. MOF-based membranes for gas separations. *Chemical Reviews* 120(16): 8161–8266. https://doi.org/10.1021/acs.chemrev.0c00119.

Radovanovic, Philip, Stephen W. Thiel and Sun-Tak Hwang. 1992a. Formation of asymmetric polysulfone membranes by immersion precipitation. Part I. Modelling mass transport during gelation. *Journal of Membrane Science* 65(3): 213–29. http://www.sciencedirect.com/science/article/pii/037673889287024R.

Radovanovic, P., S.W. Thiel and S.-T. Hwang. 1992b. Formation of asymmetric polysulfone membranes by immersion precipitation. Part II. The effects of casting solution and gelation bath compositions on membrane structure and skin formation. *Journal of Membrane Science* 65(3): 231–46. http://www.sciencedirect.com/science/article/pii/037673889287025S.

Raju, Muralikrishna, Daniel T. Banuti, Peter C. Ma and Matthias Ihme. 2017. Widom lines in binary mixtures of supercritical fluids. *Scientific Reports* 7(1): 3027. https://doi.org/10.1038/s41598-017-03334-3.

Rance, G.A. and A.N. Khlobystov. 2014. Interactions of carbon nanotubes and gold nanoparticles: the effects of solvent dielectric constant and temperature on controlled assembly of superstructures. *Dalton Transactions* 43(20): 7400–7406. http://dx.doi.org/10.1039/C3DT53372G.

Rao, K. Venkata and Subi J. George. 2010. Synthesis and controllable self-assembly of a novel coronene bisimide amphiphile. *Organic Letters* 12(11): 2656–59. https://doi.org/10.1021/ol100864e.

Ravi, R. and V. Guruprasad. 2008. Lennard-jones fluid and diffusivity: validity of the hard-sphere model for diffusion in simple fluids and application to CO_2. *Industrial & Engineering Chemistry Research* 47(4): 1297–1303. https://doi.org/10.1021/ie071073v.

Reichardt, Christian. 1994. Solvatochromic dyes as solvent polarity indicators. *Chemical Reviews* 94(8): 2319–58. http://dx.doi.org/10.1021/cr00032a005.

Rekharsky, Mikhail and Yoshihisa Inoue. 2012. Solvation effects in supramolecular recognition. *Supramolecular Chemistry*. https://doi.org/10.1002/9780470661345.smc009.

Reuvers, A.J., J.W.A. van den Berg and C.A. Smolders. 1987. Formation of membranes by means of immersion precipitation: Part I. A model to describe mass transfer during immersion precipitation. *Journal of Membrane Science* 34(1): 45–65. http://www.sciencedirect.com/science/article/pii/S0376738800800204.

Reuvers, A.J. and C.A. Smolders. 1987. Formation of membranes by means of immersion precipitation: Part II. The mechanism of formation of membranes prepared from the system cellulose acetate-acetone-water. *Journal of Membrane Science* 34(1): 67–86. http://www.sciencedirect.com/science/article/pii/S0376738800800216.

Robertson Craig, C., S. Wright Jales, J. Carrington Elliot, N. Perutz Robin, A. Hunter Christopher and Brammer Lee. 2017. Hydrogen bonding vs. halogen bonding: the solvent decides. *Chemical Science* 8(8): 5392–98. http://dx.doi.org/10.1039/C7SC01801K.

Roccatano, Danilo. 2008. Computer simulations study of biomolecules in non-aqueous or cosolvent/water mixture solutions. *Current Protein & Peptide Science* 9(4): 407–26. http://www.eurekaselect.com/node/83031/article.

Rodenburg, Jeroen, Marjolein Dijkstra and René van Roij. 2017. Van't Hoff's law for active suspensions: the role of the solvent chemical potential. *Soft Matter* 13(47): 8957–63. http://dx.doi.org/10.1039/C7SM01432E.

Rodriguez-Laguna, M.R., A. Castro-Alvarez, M. Sledzinska, J. Maire, F. Costanzo, B. Ensing et al. 2018. Mechanisms behind the enhancement of thermal properties of graphene nanofluids. *Nanoscale* 10: 15402–15409.

Saal, Christoph and Anita Nair. 2020. *Solubility in Pharmaceutical Chemistry*. Berlin, Boston: De Gruyter. https://www.degruyter.com/view/title/533927.

Saintillan, David. 2018. Rheology of active fluids. *Annual Review of Fluid Mechanics* 50(1): 563–92. https://doi.org/10.1146/annurev-fluid-010816-060049.

Saitow, Ken-ichi, Daisuke Kajiya and Keiko Nishikawa. 2004. Dynamics of density fluctuation of supercritical fluid mapped on phase diagram. *Journal of the American Chemical Society* 126(2): 422–23. https://doi.org/10.1021/ja038176z.

Sakonidou, E.P., H.R. van den Berg, C.A. ten Seldam and J.V. Sengers. 1996. The thermal conductivity of methane in the critical region. *The Journal of Chemical Physics* 105(23): 10535–55. https://doi.org/10.1063/1.472943.

Sakonidou, E.P., H.R. van den Berg, C.A. ten Seldam and J.V. Sengers. 1998. The thermal conductivity of an equimolar methane–ethane mixture in the critical region. *The Journal of Chemical Physics* 109(2): 717–36. https://doi.org/10.1063/1.476611.

Sakuma, H. and M. Ichiki. 2016. Density and isothermal compressibility of supercritical H_2O–NaCl fluid: molecular dynamics study from 673 to 2000 K, 0.2 to 2 GPa, and 0 to 22 Wt% NaCl concentrations. *Geofluids* 16(1): 89–102. https://doi.org/10.1111/gfl.12138.

Samaddar, Pallabi and Kamalika Sen. 2014. Cloud point extraction: a sustainable method of elemental preconcentration and speciation. *Journal of Industrial and Engineering Chemistry* 20(4): 1209–19. http://www.sciencedirect.com/science/article/pii/S1226086X13005133.

Samori Chhiara, Torri Cristian, Samori Giulia, Fabbri Daniele, GAlletti Paola, Guerrini Franca et al. 2010. Extraction of hydrocarbons from microalga Botryococcus Braunii with switchable solvents. *Bioresource Technology* 101(9): 3274–79. http://www.sciencedirect.com/science/article/pii/S0960852409017337.

Sánchez-Camargo, Andrea del Pilar, Jose A. Mendiola and Elena Ibáñez. 2018. CHAPTER 17 Gas expanded-liquids. pp. 512–31. *In: Supercritical and Other High-Pressure Solvent Systems: For Extraction, Reaction and Material Processing*, The Royal Society of Chemistry. http://dx.doi.org/10.1039/9781788013543-00512.

Sanchez-Fernandez Adrian, Arnold Thomas, J. Jackson Andrew, L. Fussell Sian, K. Heenan Richard, A. Campbell Richard et al. 2016. Micellization of alkyltrimethylammonium bromide surfactants in choline chloride:glycerol deep eutectic solvent. *Physical Chemistry Chemical Physics* 18(48): 33240–49. http://dx.doi.org/10.1039/C6CP06053F.

Sanchez Bruno, Calderon Cristian, Garrido Constanza, Contreras Renato and R. Campodonico Paola. 2018. Solvent effect on a model SNAr reaction in ionic liquid/water mixtures at different compositions. *New Journal of Chemistry* 42(12): 9645–50. http://dx.doi.org/10.1039/C7NJ04820C.

Saunders, Brian R., Helen M. Crowther and Brian Vincent. 1997. Poly[(methyl methacrylate)-co-(methacrylic acid)] microgel particles: swelling control using PH, cononsolvency, and osmotic deswelling. *Macromolecules* 30(3): 482–87. https://doi.org/10.1021/ma961277f.

Schein, Catherine H. 1990. Solubility as a function of protein structure and solvent components. *Bio/Technology* 8(4): 308–17. https://doi.org/10.1038/nbt0490-308.

Schellman, John A. 1987. Selective binding and solvent denaturation. *Biopolymers* 26(4): 549–59. https://doi.org/10.1002/bip.360260408.

Schellman, J.A. 1990. A simple model for solvation in mixed solvents: applications to the stabilization and destabilization of macromolecular structures. *Biophysical Chemistry* 37(1): 121–40. https://www.sciencedirect.com/science/article/pii/030146229088013I.

Schellman, J.A. 1994. The thermodynamics of solvent exchange. *Biopolymers* 34(8): 1015–26. https://doi.org/10.1002/bip.360340805.

Schmitt, William J. and Robert C. Reid. 1986. The use of entrainers in modifying the solubility of phenanthrene and benzoic acid in supercritical carbon dioxide and ethane. *Fluid Phase Equilibria* 32(1): 77–99. http://www.sciencedirect.com/science/article/pii/0378381286870078.

Schuur, Boelo, Thomas Brouwer, Dion Smink and Lisette M.J. Sprakel. 2019. Green solvents for sustainable separation processes. *Current Opinion in Green and Sustainable Chemistry* 18: 57–65. http://www.sciencedirect.com/science/article/pii/S2452223618300919.

Schuur, Boelo, Mart Nijland, Marek Blahušiak and Alberto Juan. 2018. CO_2-switchable solvents as entrainer in fluid separations. *ACS Sustainable Chemistry & Engineering* 6(8): 10429–35. https://doi.org/10.1021/acssuschemeng.8b01771.

Scruggs, Neal R., Julia A. Kornfield and Jyotsana Lal. 2006. Using the 'Switchable' quality of liquid crystal solvents to mediate segregation between coil and liquid crystalline polymers. *Macromolecules* 39(11): 3921–26. https://doi.org/10.1021/ma052414o.

Sels, Hannes, Herwig De Smet and Jeroen Geuens. 2020. SUSSOL—using artificial intelligence for greener solvent selection and substitution. *Molecules* 25(13).

Sheldon, Roger A. 2019. The greening of solvents: towards sustainable organic synthesis. *Current Opinion in Green and Sustainable Chemistry* 18: 13–19. http://www.sciencedirect.com/science/article/pii/S2452223618301172.

Shen, S.S., K.P. Liu, J.J. Yang, Y. Li, R.B. Bai and X.J. Zhou. 2019. Application of a triblock copolymer additive modified polyvinylidene fluoride membrane for effective oil/water separation. *Royal Society Open Science* 5(5): 171979. https://doi.org/10.1098/rsos.171979.

Shi Lei, Hao Huiying, Dong Jingjing, Zhong Tingting, Zhang Chen, Hao Jiabin et al. 2019. Solvent engineering for intermediates phase, all-ambient-air-processed in organic–inorganic hybrid perovskite solar cells. *Nanomaterials* 9(7): 915.

Shi, Zhengqi and Ahalapitiya H. Jayatissa. 2018. Perovskites-based solar cells: a review of recent progress, materials and processing methods. *Materials (Basel, Switzerland)* 11(5): 729. https://www.ncbi.nlm.nih.gov/pubmed/29734667.

Shuai Li, Amiri Masoud Talebi, Questell-Santiago Ydna M., Héroguel Florent, Li Yanding, Kim Hoon et al. 2016. Formaldehyde stabilization facilitates lignin monomer production during biomass depolymerization. *Science* 354(6310): 329–33. https://doi.org/10.1126/science.aaf7810.

Silva, Maria Fernanda, Estela Soledad Cerutti and Luis D. Martinez. 2006. Coupling cloud point extraction to instrumental detection systems for metal analysis. *Microchimica Acta* 155(3): 349. https://doi.org/10.1007/s00604-006-0511-3.

Simeoini, G.G., T. Bryk, F.A. Gorelli, M. Krisch, G. Ruocco, M. Santoro et al. 2010. The Widom line as the crossover between liquid-like and gas-like behaviour in supercritical fluids. *Nature Physics* 6: 503. https://doi.org/10.1038/nphys1683.

Singh Amarpreet, Walvekar Rashmi, Khalid Mohammad, Wong Wai Yin and T.C.S.M. Gupta. 2018. Thermophysical properties of glycerol and polyethylene glycol (PEG 600) based DES. *Journal of Molecular Liquids* 252: 439–44. http://www.sciencedirect.com/science/article/pii/S0167732217328647.

Siougkrou, Eirini, Amparo Galindo and Claire S. Adjiman. 2014. On the optimal design of gas-expanded liquids based on process performance. *Chemical Engineering Science* 115: 19–30. http://www.sciencedirect.com/science/article/pii/S0009250913008208.

Skouta, Rachid. 2009. Selective chemical reactions in supercritical carbon dioxide, water, and ionic liquids. *Green Chemistry Letters and Reviews* 2(3): 121–56. https://doi.org/10.1080/17518250903230001.

Slakman, Belinda L. and Richard H. West. 2019. Kinetic solvent effects in organic reactions. *Journal of Physical Organic Chemistry* 32(3): e3904. https://doi.org/10.1002/poc.3904.

Sleczkowski Marcin, L., F.J. Mabesoone Mathijs, Sleczkowski Piotr, R.A. Palmans Anja and E.W. Meijer. 2021. Competition between chiral solvents and chiral monomers in the helical bias of supramolecular polymers. *Nature Chemistry* 13(2): 200–207. https://doi.org/10.1038/s41557-020-00583-0.

Smolders, C.A., A.J. Reuvers, R.M. Boom and I.M. Wienk. 1992. Microstructures in phase-inversion membranes. Part 1. Formation of macrovoids. *Journal of Membrane Science* 73(2): 259–75. http://www.sciencedirect.com/science/article/pii/0376738892801346.

Snezhko, Alexey and Aranson, Igor S. 2011. Magnetic manipulation of self-assembled colloidal asters. *Nature Materials* 10(9): 698–703. https://doi.org/10.1038/nmat3083.

Lu, B. C.-Y., D. Zhang and W. Sheng. 1990. Solubility enhancement in supercritical solvents. *Pure and Applied Chemistry* 62: 2277. https://www.degruyter.com/view/j/pac.1990.62.issue-12/pac199062122277/pac199062122277.xml.

Song, Di, A. Frank Seibert and Gary T. Rochelle. 2018. Mass transfer parameters for packings: effect of viscosity. *Industrial & Engineering Chemistry Research* 57(2): 718–29. https://doi.org/10.1021/acs.iecr.7b04396.

Sovová Helena and Stateva Roumiana P. 2011. Supercritical fluid extraction from vegetable materials. *Reviews in Chemical Engineering* 27: 79. https://www.degruyter.com/view/j/revce.2011.27.issue-3-4/revce.2011.002/revce.2011.002.xml.

Span, Roland and Wolfgang Wagner. 1996. A new equation of state for carbon dioxide covering the fluid region from the triple-point temperature to 1100 K at pressures up to 800 MPa. *Journal of Physical and Chemical Reference Data* 25(6): 1509–96. https://doi.org/10.1063/1.555991.

Spange, Stefan and Nadine Weiß. 2023. Empirical hydrogen bonding donor (HBD) parameters of organic solvents using solvatochromic probes—a critical evaluation. *ChemPhysChem* 24(9): e202200780. https://doi.org/10.1002/cphc.202200780.

Stalikas, Constantine D. 2002. Micelle-mediated extraction as a tool for separation and preconcentration in metal analysis. *TrAC Trends in Analytical Chemistry* 21(5): 343–55. http://www.sciencedirect.com/science/article/pii/S0165993602005022.

Stavrou, Marina, Matthias Lampe, André Bardow and Joachim Gross. 2014. Continuous molecular targeting–computer-aided molecular design (CoMT–CAMD) for simultaneous process and solvent design for CO_2 capture. *Industrial & Engineering Chemistry Research* 53(46): 18029–41. https://doi.org/10.1021/ie502924h.

Stepankova Veronika, Bidmanova Sarka, Koudelakova Tana, Prokop Zbynek, Chaloupkova Radka and Damborsky Jiri. 2013. Strategies for stabilization of enzymes in organic solvents. *ACS Catalysis* 3(12): 2823–36. https://doi.org/10.1021/cs400684x.

Stewart, M. and K. Arnold. 2008. *Gas-Liquid And Liquid-Liquid Separators*. Elsevier Science. https://books.google.fr/books?id=gT4k2mFkjL8C.

Strathmann, H., K. Kock, P. Amar and R.W. Baker. 1975. The formation mechanism of asymmetric membranes. *Desalination* 16(2): 179–203. http://www.sciencedirect.com/science/article/pii/S0011916400820925.

Subbotina Elena, Rukkijakan Thanya, M. Marquez-Medina Dolores, Yu Xiaowen, Johnsson Mats and S.M. Samec Joseph. 2021. Oxidative cleavage of C–C bonds in lignin. *Nature Chemistry* 13(11): 1118–25. https://doi.org/10.1038/s41557-021-00783-2.

Subramaniam, Bala. 2010a. Exploiting neoteric solvents for sustainable catalysis and reaction engineering: opportunities and challenges. *Industrial & Engineering Chemistry Research* 49(21): 10218–29. http://pubs.acs.org/doi/abs/10.1021/ie101543a.

Subramaniam, B. 2010b. Gas-expanded liquids for sustainable catalysis and novel materials: recent advances. *Coordination Chemistry Reviews* 254(15–16): 1843–53. http://www.sciencedirect.com/science/article/pii/S0010854509003245.

Subramaniam, B. 2012. Gas Expanded Liquids (GXL) for sustainable catalysis - encyclopedia of sustainability science and technology. pp. 3933–55. *In*: Robert A. Meyers (ed.). New York, NY: Springer New York. https://doi.org/10.1007/978-1-4419-0851-3_328.

Subramaniam, B., R.V. Chaudhari, A.S. Chaudhari, G.R. Akien and Z. Xie. 2014. Supercritical fluids and gas-expanded liquids as tunable media for multiphase catalytic reactions. *Chemical Engineering Science* 115: 3–18. http://www.sciencedirect.com/science/article/pii/S0009250914001122.

Suisse Jean-Moïse, Bellemin-Laponnaz Stéphane, Douce Laurent, François Aline-Maisse and Walter Richard. 2005. A new liquid crystal compound based on an ionic imidazolium salt. *Tetrahedron Letters* 46(25): 4303–5. https://www.sciencedirect.com/science/article/pii/S0040403905009536.

Taft, R.W., Thor Gramstad and Mortimer J. Kamlet. 1982. Linear solvation energy relationships. 14. Additions to and correlations with the β Scale of hydrogen bond acceptor basicities. *The Journal of Organic Chemistry* 47(23): 4557–63. https://doi.org/10.1021/jo00144a030.

Taft, R.W. and Mortimer J. Kamlet. 1976. The solvatochromic comparison method. 2. The α-scale of solvent hydrogen-bond donor (HBD) acidities. *Journal of the American Chemical Society* 98(10): 2886–94. https://doi.org/10.1021/ja00426a036.

Taft, Robert W., Michael H. Abraham, Ruth M. Doherty and Mortimer J. Kamlet. 1985. Linear solvation energy relationships. 29. Solution properties of some tetraalkylammonium halide ion pairs and dissociated ions. *Journal of the American Chemical Society* 107(11): 3105–10. https://doi.org/10.1021/ja00297a016.

Takatori, Sho C. and John F. Brady. 2015. A theory for the phase behavior of mixtures of active particles. *Soft Matter* 11(40): 7920–31. http://dx.doi.org/10.1039/C5SM01792K.

Tanaka Fumihiko, Koga Tsuyoshi, Kojima Hiroyuki, Xue Na and Winnik Françoise M. 2011. Preferential adsorption and co-nonsolvency of thermoresponsive polymers in mixed solvents of water/methanol. *Macromolecules* 44(8): 2978–89. https://doi.org/10.1021/ma102695n.

Tanaka, Fumihiko, Tsuyoshi Koga and Françoise M. Winnik. 2008. Temperature-responsive polymers in mixed solvents: competitive hydrogen bonds cause cononsolvency. *Physical Review Letters* 101(2): 28302. https://link.aps.org/doi/10.1103/PhysRevLett.101.028302.

Tapia, O. 1992. Solvent effect theories: quantum and classical formalisms and their applications in chemistry and biochemistry. *Journal of Mathematical Chemistry* 10(1): 139–81. https://doi.org/10.1007/BF01169173.

Thornton, J.D. 1992. *Science and Practice of Liquid-Liquid Extraction: Phase Equilibria, Mass Transfer and Interfacial Phenomena, Extractor Hydrodynamics, Selection, and Design.* Clarendon Press. https://books.google.fr/books?id=IKJTAAAAMAAJ.

Thutupalli Shashi, Geyer Delphine, Singh Rajesh and A. Stone Howard. 2018. Flow-induced phase separation of active particles is controlled by boundary conditions. *Proceedings of the National Academy of Sciences* 115(21): 5403–8. https://doi.org/10.1073/pnas.1718807115.

Tilly, Kevin D., Neil R. Foster, Stuart J. Macnaughton and David L. Tomasko. 1994. Viscosity correlations for binary supercritical fluids. *Industrial & Engineering Chemistry Research* 33(3): 681–88. https://doi.org/10.1021/ie00027a028.

Timasheff, Serge N. 2002. Protein-solvent preferential interactions, protein hydration, and the modulation of biochemical reactions by solvent components. *Proceedings of the National Academy of Sciences* 99(15): 9721 LP–9726. http://www.pnas.org/content/99/15/9721.abstract.

Tisza, L. 1942. Supersonic absorption and stokes' viscosity relation. *Physical Review* 61(7–8): 531–36. https://link.aps.org/doi/10.1103/PhysRev.61.531.

Tomich Anton W., Chen Jianjun, Carta Veronica, Guo Juchen and Lavallo Vincent. 2024. Electrolyte engineering with carboranes for next-generation Mg batteries. *ACS Central Science.* https://doi.org/10.1021/acscentsci.3c01176.

Toropova Alla P., A. Toropov Andrey, Benfanati Emilio, Gini Giuseppina, Leszczynska and Leszczynski. 2011. CORAL: QSPR models for solubility of [C60] and [C70] fullerene derivatives. *Molecular Diversity* 15(1): 249–56. https://doi.org/10.1007/s11030-010-9245-6.

Toropova, Alla P., Andrey A. Toropov and Emilio Benfenati. 2019. QSPR as a random event: solubility of fullerenes C[60] and C[70]. *Fullerenes, Nanotubes and Carbon Nanostructures* 27(10): 816–21. https://doi.org/10.1080/1536383X.2019.1649659.

Tournier Alexandre, Huang Danzhi, Schwarzl Sonja, Fischer Stefan and Smith Jeremy. 2003. Time-resolved computational protein biochemistry: solvent effects on interactions, conformational transitions and equilibrium fluctuations. *Faraday Discussions* 122(0): 243–51. http://dx.doi.org/10.1039/B201191C.

Troshin Pavel, A., Hoppe Harald, Renz Joachim, Egginger Martin, Mayorova Julia Yu, E. Goryachev Andrey et al. 2009. Material solubility-photovoltaic performance relationship in the design of novel fullerene derivatives for bulk heterojunction solar cells. *Advanced Functional Materials* 19(5): 779–88. https://doi.org/10.1002/adfm.200801189.

Tsai, J.T., Y.S. Su, D.M. Wang, J.L. Kuo, J.Y. Lai and A. Deratani. 2010. Retainment of pore connectivity in membranes prepared with vapor-induced phase separation. *Journal of Membrane Science* 362(1): 360–73. http://www.sciencedirect.com/science/article/pii/S0376738810005065.

Tsay, C.S. and A.J. Mchugh. 1990. Mass transfer modeling of asymmetric membrane formation by phase inversion. *Journal of Polymer Science Part B: Polymer Physics* 28(8): 1327–65. https://doi.org/10.1002/polb.1990.090280810.

Tsuzuki Seiji, Uchimaru Tadafumi, Mikami Masuhiro, Tanabe Kazutoshi, Sako Takeshi and Kuwajima Satoru. 1996. Molecular dynamics simulation of supercritical carbon dioxide fluid with the model potential from ab initio molecular orbital calculations. *Chemical Physics Letters* 255(4): 347–49. http://www.sciencedirect.com/science/article/pii/0009261496003971.

Tu Yongguang, Wu Jihuai, He Xin, Guo Panfeng, Wu Tongyue, Luo Hui et al. 2017. Solvent engineering for forming stonehenge-like pbi2 nano-structures towards efficient perovskite solar cells. *Journal of Materials Chemistry A* 5(9): 4376–83. http://dx.doi.org/10.1039/C6TA11004E.

Tucker, Susan C. 1999. Solvent density inhomogeneities in supercritical fluids. *Chemical Reviews* 99(2): 391–418. https://doi.org/10.1021/cr9700437.

Uragmi, Tadashi, Hiroshi Yamada and Takashi Miyata. 2006. Effects of fluorine-containing graft and block copolymer additives on removal characteristics of dilute benzene in water by microphase-separated membranes modified with these additives. *Macromolecules* 39(5): 1890–97. https://doi.org/10.1021/ma052302x.

Urík, Ingrid Hagarová and Martin. 2016. New approaches to the cloud point extraction: utilizable for separation and preconcentration of trace metals. *Current Analytical Chemistry* 12(2): 87–93. http://www.eurekaselect.com/node/131854/article.

Urkiaga, A., D. Iturbe and J. Etxebarria. 2015. Effect of different additives on the fabrication of hydrophilic polysulfone ultrafiltration membranes. *Desalination and Water Treatment* 56(13): 3415–26. https://doi.org/10.1080/19443994.2014.1000976.

Valkenburg, Michael E. Van, Robert L. Vaughn, Margaret Williams and John S. Wilkes. 2005. Thermochemistry of ionic liquid heat-transfer fluids. *Thermochimica Acta* 425(1): 181–88. http://www.sciencedirect.com/science/article/pii/S004060310400509X.

Varghese, Jithin John and Samir H. Mushrif. 2019. Origins of complex solvent effects on chemical reactivity and computational tools to investigate them: a review. *Reaction Chemistry & Engineering* 4(2): 165–206. http://dx.doi.org/10.1039/C8RE00226F.

Varma, Akhil, Thomas D. Montenegro-Johnson and Sébastien Michelin. 2018. Clustering-induced self-propulsion of isotropic autophoretic particles. *Soft Matter* 14(35): 7155–73. http://dx.doi. org/10.1039/C8SM00690C.

Veit Max, Kumar Jain Sandeep, Bonakala Satyanayana, Rudra Indranil, Hohl Detlef and Csanyi Gabor. 2019. Equation of state of fluid methane from first principles with machine learning potentials. *Journal of Chemical Theory and Computation* 15(4): 2574–86. https://doi.org/10.1021/acs. jctc.8b01242.

Veronika R. Meyer. 2010. *Practical High-Performance Liquid Chromatography*. 5th ed. John Wiley & Sons, Ltd. https://onlinelibrary.wiley.com/doi/book/10.1002/9780470688427.

Vesovic, V., M.J. Assael and Z.A. Gallis. 1998. Prediction of the viscosity of supercritical fluid mixtures. *International Journal of Thermophysics* 19(5): 1297–1313. https://doi. org/10.1023/A:1021971232631.

Walker Theodore, W., K. Chew Alex, Li Huixiang, Demir Benginur, Zhang Z. Conrad et al. 2018. Universal kinetic solvent effects in acid-catalyzed reactions of biomass-derived oxygenates. *Energy & Environmental Science* 11(3): 617–28. http://dx.doi.org/10.1039/C7EE03432F.

Walsh, John M., George D. Ikonomou and Marc D. Donohue. 1987. Supercritical phase behavior: the entrainer effect. *Fluid Phase Equilibria* 33(3): 295–314. http://www.sciencedirect.com/science/ article/pii/0378381287850422.

Walter, Jonathan, Jan Sehrt, Jadran Vrabec and Hans Hasse. 2012. Molecular dynamics and experimental study of conformation change of poly(N-isopropylacrylamide) hydrogels in mixtures of water and methanol. *The Journal of Physical Chemistry B* 116(17): 5251–59. https://doi.org/10.1021/jp212357n.

Wang Dechao, Xin Yangyang, Yao Dongdong, Li Xiaoqian, NongHailong, Zhang Hongmin et al. 2022. Shining light on porous liquids: from fundamentals to syntheses, applications and future challenges. *Advanced Functional Materials* 32(1): 2104162. https://doi.org/10.1002/ adfm.202104162.

Wang Fei, Zhou Xianfang, Liang Xiao, Duan Dawei, Ge Chuang-Ye, Lin Haoran et al. 2023. Solvent engineering of ionic liquids for stable and efficient perovskite solar cells. *Advanced Energy and Sustainability Research* 4(1): 2200140. https://doi.org/10.1002/aesr.202200140.

Wang Haoliang, Deng Lianliang, Pan Yiyi, Zhang Xin, Li Xiaoguo, Wang Yanyan et al. 2023. Green solvent polishing enables highly efficient quasi-2D perovskite solar cells. *ACS Applied Materials & Interfaces* 15(30): 36447–56. https://doi.org/10.1021/acsami.3c08182.

Wang, Hongwu and Arieh Ben-Naim. 1997. Solvation and solubility of globular proteins. *The Journal of Physical Chemistry B* 101(6): 1077–86. https://doi.org/10.1021/jp961591b.

Wang, Jiayuan and Richard Lakerveld. 2018a. Integrated solvent and process design for continuous crystallization and solvent recycling using PC-SAFT. *AIChE Journal* 64(4): 1205–16. https://doi.org/10.1002/aic.15998.

Wang, J. and R. Lakerveld. 2018b. Integrated solvent and process optimization using PC-SAFT for continuous crystallization with energy-intensive solvent separation for recycling. pp. 1051–56. *In*: Mario R. Eden, Marianthi G. Ierapetritou, and Gavin P.B.T. (eds.). *13 International Symposium on Process Systems Engineering (PSE 2018)* - Computer Aided Chemical Engineering Towler. Elsevier. https://www.sciencedirect.com/science/article/pii/B9780444642417501701.

Wang, Jiayuan, Lingyu Zhu and Richard Lakerveld. 2020. A hybrid framework for simultaneous process and solvent optimization of continuous anti-solvent crystallization with distillation for solvent recycling. *Processes* 8(1).

Wang Mengmeng, Liu Kang, Xu Zibo, Dutta Shanta, Valix Marjorie, S. Alessi Daniel et al. 2023. Selective extraction of critical metals from spent lithium-ion batteries. *Environmental Science & Technology* 57(9): 3940–50. https://doi.org/10.1021/acs.est.2c07689.

Wang, S., X. Meng, H. Zhou, Y. Liu, F. Secundo and Y. Liu. 2016. Enzyme stability and activity in non-aqueous reaction systems: a mini review. *Catalysts* 6(2).

Wang, Zhiyuan, Baojiang Sun and Linlin Yan. 2015. Improved density correlation for supercritical CO_2. *Chemical Engineering & Technology* 38(1): 75–84. https://doi.org/10.1002/ceat.201400357.

Wangler, Anton, Christoph Held and Gabriele Sadowski. 2019. Thermodynamic activity-based solvent design for bioreactions. *Trends in Biotechnology* 37(10): 1038–41. https://www.sciencedirect.com/science/article/pii/S0167779919301076.

Wei, Ming, Ghezai T. Musie, Daryle H. Busch and Bala Subramaniam. 2002. CO_2-expanded solvents: unique and versatile media for performing homogeneous catalytic oxidations. *Journal of the American Chemical Society* 124(11): 2513–17. http://dx.doi.org/10.1021/ja0114411.

Welton, Tom. 2015. Solvents and sustainable chemistry. *Proceedings of the Royal Society A: Mathematical, Physical and Engineering Sciences* 471(2183): 20150502. https://doi.org/10.1098/rspa.2015.0502.

Weng, Caihong, Xiaowei Peng and Yejun Han. 2021. Depolymerization and conversion of lignin to value-added bioproducts by microbial and enzymatic catalysis. *Biotechnology for Biofuels* 14(1): 84. https://doi.org/10.1186/s13068-021-01934-w.

Wensink, H.H., V. Kantsler, R.E. Goldstein and J. Dunkel. 2014. Controlling active self-assembly through broken particle-shape symmetry. *Physical Review E* 89(1): 10302. https://link.aps.org/doi/10.1103/PhysRevE.89.010302.

Wertheim, M.S. 1984. Fluids with highly directional attractive forces. I. Statistical thermodynamics. *Journal of Statistical Physics* 35(1): 19–34. https://doi.org/10.1007/BF01017362.

Wesch, A., N. Dahmen and K.H. Ebert. 1996. Measuring the static dielectric constants of pure carbon dioxide and carbon dioxide mixed with ethanol and toluene at elevated pressures. *Berichte der Bunsengesellschaft für physikalische Chemie* 100(8): 1368–71. https://doi.org/10.1002/bbpc.19961000816.

Wescott Charles R. and Klibanov Alexander M. 1994. The solvent dependence of enzyme specificity. Biochimica et Biophysica Acta (BBA) - Protein Structure and Molecular Enzymology 1206(1): 1–9.

Wijmans, J.G., J. Kant, M.H.V. Mulder and C.A. Smolders. 1985. Phase separation phenomena in solutions of polysulfone in mixtures of a solvent and a nonsolvent: relationship with membrane formation. *Polymer* 26(10): 1539–45. http://www.sciencedirect.com/science/article/pii/0032386185900904.

Wilhelm, E. 2007. Thermodynamics of nonelectrolyte solubility in development and applications in solubility, T.M. Letcher Ed. The Royal Society of Chemistry. *Developments and Applications in Solubility*. http://dx.doi.org/10.1039/9781847557681.

William, J. Leigh and Mark, S. Workentin. 1998. Liquid crystals as solvents for spectroscopic, chemical reaction, and gas chromatographic applications. pp. 839–95. *In*: Vill, V., D. Demus,

J. Goodby, G. W. Gray and H.-W. Spiess (eds.). *Handbook of Liquid Crystals.* Weinheim: WILEY-VCH Verlag GmbH.

Williams, Robert O. (Bill). 2000. Solubility and Solubilization in Aqueous Media By Samuel H. Yalkowsky (University of Arizona). Oxford University Press: New York. 1999. ISBN 0-8412-3576-7. *Journal of the American Chemical Society* 122(40): 9882. https://doi.org/10.1021/ja0047424.

Wingren, C. and U.-B. Hansson. 2000. CHROMATOGRAPHY: LIQUID | Partition Chromatography (Liquid–Liquid). pp. 760–70. *In*: Ian, D.B.T. (ed.). Encyclopedia of Separation Science Wilson. Oxford: Academic Press. http://www.sciencedirect.com/science/article/pii/B0122267702018019.

Winnik, F.M., M.F. Ottaviani, S.H. Bossmann, M. Garcia-Garibay and N.T. Turro. 1992. Consolvency of poly(N-isopropylacrylamide) in mixed water-methanol solutions: a look at spin-labeled polymers. *Macromolecules* 25(22): 6007–17. https://doi.org/10.1021/ma00048a023.

van de Witte, P., P.J. Dijkstra, J.W.A. van den Berg and J. Feijen. 1996. Phase separation processes in polymer solutions in relation to membrane formation. *Journal of Membrane Science* 117(1): 1–31. http://www.sciencedirect.com/science/article/pii/0376738896000889.

Wu, J., J. Li, M. Dong, K. Miao, Y. Miao, Y. Wu et al. 2018. Solvent effect on host–guest two-dimensional self-assembly mediated by halogen bonding. *The Journal of Physical Chemistry C* 122(39): 22597–604. https://doi.org/10.1021/acs.jpcc.8b07194.

Wu, X., T. Liu, Y.F. Bai, M.M. Rahman, Ch-J. Sun et al. 2020. Effects of solvent formulations in electrolytes on fast charging of li-ion cells. *Electrochimica Acta* 353: 136453. https://www.sciencedirect.com/science/article/pii/S001346862030846X.

Wu, X., Y. Zheng, J. Liang, Z. Zhang, C. Tian, Z. Zhang et al. 2023. Green-solvent-processed formamidinium-based perovskite solar cells with uniform grain growth and strengthened interfacial contact via a nanostructured tin oxide layer. *Materials Horizons* 10(1): 122–35. http://dx.doi.org/10.1039/D2MH00970F.

Wu, Y., J. Li, Y. Yuan, M. Dong, B. Zha, X. Miao et al. 2017. Halogen bonding versus hydrogen bonding induced 2D self-assembled nanostructures at the liquid–solid interface revealed by STM. *Physical Chemistry Chemical Physics* 19(4): 3143–50. http://dx.doi.org/10.1039/C6CP08054E.

Xie, Zhuanzhuan and Bala Subramaniam. 2014. Development of a greener hydroformylation process guided by quantitative sustainability assessments. *ACS Sustainable Chemistry & Engineering* 2(12): 2748–57. https://doi.org/10.1021/sc500483f.

Xu, Zhijie, Rajesh Singh, Jie Bao and Chao Wang. 2019. *Direct Effect of Solvent Viscosity on the Physical Mass Transfer for Wavy Film Flow in a Packed Column.*

Xue, S., P. Xing, J. Zhang, Y. Zeng and Y. Zhao. 2019. Diverse role of solvents in controlling supramolecular chirality. *Chemistry – A European Journal* 25(31): 7426–37. https://doi.org/10.1002/chem.201900714.

Y., Marcus. 2017. Preferential solvation of drugs in binary solvent mixtures. *Pharmaceutica Analytica Acta* 8(1): 1000537. DOI: 10.4172/2153-2435.1000537.

Yadav, Anita and Siddharth Pandey. 2014. Densities and viscosities of (choline chloride + urea) deep eutectic solvent and its aqueous mixtures in the temperature range 293.15 K to 363.15 K. *Journal of Chemical & Engineering Data* 59(7): 2221–29. https://doi.org/10.1021/je5001796.

Yamamoto, Takeshi et al. 2018. Self-assembly of nanocubic molecular capsules via solvent-guided formation of rectangular blocks. *The Journal of Physical Chemistry Letters* 9(20): 6082–88. https://doi.org/10.1021/acs.jpclett.8b02624.

Yan, Q., A. Ding, M. Li, C. Liu and C. Xiao. 2023. Green leaching of lithium-ion battery cathodes by ascorbic acid modified guanidine-based deep eutectic solvents. *Energy & Fuels* 37(2): 1216–24. https://doi.org/10.1021/acs.energyfuels.2c03699.

Yang, Mingcheng, Adam Wysocki and Marisol Ripoll. 2014. Hydrodynamic simulations of self-phoretic microswimmers. *Soft Matter* 10(33): 6208–18. http://dx.doi.org/10.1039/C4SM00621F.

Yao, J., D.B. Lao, X. Sui, Y. Zhou, S.K. Nune, X. Ma et al. 2017. Two coexisting liquid phases in switchable ionic liquids. *Physical Chemistry Chemical Physics* 19(34): 22627–32. http://dx.doi.org/10.1039/C7CP03754F.

Yao, X., D. Collin, O. Gavat, A. Carvalho, E. Moulin, N. Giuseppone et al. 2022. Effect of solvent isomers on the gelation properties of tri-aryl amine organogels and their hybrid thermoreversible gels with poly[vinyl chloride]. *Soft Matter* 18(30): 5575–84. http://dx.doi.org/10.1039/D2SM00563H.

Yeow, M.L., Y.T. Liu and K. Li. 2004. Morphological study of poly(vinylidene fluoride) asymmetric membranes: effects of the solvent, additive, and dope temperature. *Journal of Applied Polymer Science* 92(3): 1782–89. https://doi.org/10.1002/app.20141.

Yerushalmi, R., A. Brandis, V. Rosenbach-Belkin, K.K. Baldridge and A. Scherz. 2006. Modulation of fragmental charge transfer via hydrogen bonds. direct measurement of electronic contributions. *The Journal of Physical Chemistry A* 110(2): 412–21. https://doi.org/10.1021/jp052809+.

Yokoyama, Tai, R.W. Taft and Mortimer J. Kamlet. 1976. The solvatochromic comparison method. 3. Hydrogen bonding by some 2-nitroaniline derivatives. *Journal of the American Chemical Society* 98(11): 3233–37. https://doi.org/10.1021/ja00427a030.

Yoo, Ji-Hyun, Alexander Breitholz, Yoshio Iwai and Ki-Pung Yoo. 2012. Diffusion coefficients of supercritical carbon dioxide and its mixtures using molecular dynamic simulations. *Korean Journal of Chemical Engineering* 29(7): 935–40. https://doi.org/10.1007/s11814-011-0248-5.

Yu, X., D. Gao, Z. Li, X. Sun, B. Li, Z. Zhu et al. 2023. Green-solvent processable dopant-free hole transporting materials for inverted perovskite solar cells. *Angewandte Chemie International Edition* 62(11): e202218752. https://doi.org/10.1002/anie.202218752.

Yu, Y., M. Cirelli, B. Kieviet, E.S. Kooji, G.J. Vancso and S. de Beer. 2016. Tunable friction by employment of co-non-solvency of PNIPAM brushes. *Polymer* 102: 372–78. http://www.sciencedirect.com/science/article/pii/S0032386116306863.

Yun, S.L. Jimmy, Angela K. Dillow and Charles A. Eckert. 1996. Density measurements of binary supercritical fluid ethane/cosolvent mixtures. *Journal of Chemical & Engineering Data* 41(4): 791–93. https://doi.org/10.1021/je960032v.

Zabaloy, M.S., J.M.V. Machado and E.A. Macedo. 2001. A study of lennard–jones equivalent analytical relationships for modeling viscosities. *International Journal of Thermophysics* 22(3): 829–58. https://doi.org/10.1023/A:1010779000264.

Zabaloy, Marcelo S., Victor R. Vasquez and Eugénia A. Macedo. 2005. Viscosity of pure supercritical fluids. *The Journal of Supercritical Fluids* 36(2): 106–17. http://www.sciencedirect.com/science/article/pii/S0896844605001063.

Zhang, Chaofeng and Feng Wang. 2020. Catalytic lignin depolymerization to aromatic chemicals. *Accounts of Chemical Research* 53(2): 470–84. https://doi.org/10.1021/acs.accounts.9b00573.

Zhang, C., Z. Song, C. Jin, T. Zhou, T. Noël, H. Gröger et al. 2020. Screening of functional solvent system for automatic aldehyde and ketone separation in aldol reaction: a combined COSMO-RS and experimental approach. *Chemical Engineering Journal* 385: 123399. http://www.sciencedirect.com/science/article/pii/S1385894719328128.

Zhang, H., K. Darabi, N.Y. Nia, A. Krishna, P. Ahlawat, B. Guo et al. 2022. A universal co-solvent dilution strategy enables facile and cost-effective fabrication of perovskite photovoltaics. *Nature Communications* 13(1): 89. https://doi.org/10.1038/s41467-021-27740-4.

Zhang, Jianguo and Bo Hu. 2013. Liquid-liquid extraction (LLE). *Separation and Purification Technologies in Biorefineries*: 61–78. https://doi.org/10.1002/9781118493441.ch3.

Zhang, J., L. Zhang, X. Li, X. Zhu, J. Yu and K. Fan. 2019. Binary solvent engineering for high-performance two-dimensional perovskite solar cells. *ACS Sustainable Chemistry & Engineering* 7(3): 3487–95. https://doi.org/10.1021/acssuschemeng.8b05734.

Zhang, S., S. Wu, W. Chen, H. Zhu, Z. Xiong, Z. Yang et al. 2018. Solvent engineering for efficient inverted perovskite solar cells based on inorganic CsPbI2Br light absorber. *Materials Today Energy* 8: 125–33. http://www.sciencedirect.com/science/article/pii/S246860691830008X.

Zhang, Y., F. Wang, W. Zhang, S. Ren, Y. Hou and W. Wu. 2024. High-selectivity recycling of valuable metals from spent lithium-ion batteries using recyclable deep eutectic solvents. *ChemSusChem* n/a(n/a): e202301774. https://doi.org/10.1002/cssc.202301774.

Zhang, Z., Y. Tu, Y. Hui, W. Liu, Z. Zhou and Z. Ren. 2020. Preparation and application of CO_2-triggered switchable solvents in separation of toluene/n-heptane. *Langmuir* 36(2): 510–19. https://doi.org/10.1021/acs.langmuir.9b02890.

Zhao, Y., M.K. Singh, K. Kremer, R. Cortes-Huerto and D. Mukherji. 2020. Why do elastin-like polypeptides possibly have different solvation behaviors in water–ethanol and water–urea mixtures? *Macromolecules* 53(6): 2101–10. https://doi.org/10.1021/acs.macromol.9b02123.

Zhenova, Anna. 2020. Challenges in the development of new green solvents for polymer dissolution. *Polymer International* 69(10): 895–901. https://doi.org/10.1002/pi.6072.

Zhong, N., C. Lei, R. Meng, J. Li, X. He and X. Liang. 2022. Electrolyte solvation chemistry for the solution of high-donor-number solvent for stable Li–S batteries. *Small* 18(16): 2200046. https://doi.org/10.1002/smll.202200046.

Zhou, P., W. Hou, Y. Xia, Y. Ou, H-Y. Zhou, W. Zhang et al. 2023. Tuning and balancing the donor number of lithium salts and solvents for high-performance Li metal anode. *ACS Nano* 17(17): 17169–79. https://doi.org/10.1021/acsnano.3c05016.

Zhu, Peng-wei and Luguang Chen. 2019. Effects of cosolvent partitioning on conformational transitions and chain flexibility of thermoresponsive microgels. *Physical Review E* 99(2): 22501. https://link.aps.org/doi/10.1103/PhysRevE.99.022501.

Zhuang, Quan, Bruce Clements, Junyi Dai and Logan Carrigan. 2016. Ten years of research on phase separation absorbents for carbon capture: achievements and next steps. *International Journal of Greenhouse Gas Control* 52: 449–60. http://www.sciencedirect.com/science/article/pii/S1750583616301931.

Zhuohua Sun, Fridrich Balint, de Santi Alessandra, Elangovan Saravanakumar and Barta Katalin. 2018. Bright side of lignin depolymerization: toward new platform chemicals. *Chemical Reviews* 118(2): 614–78. https://doi.org/10.1021/acs.chemrev.7b00588.

Zuo, T., C. Ma, G. Jiao, Z. Han, S. Xiao, H. Liang et al. 2019. Water/cosolvent attraction induced phase separation: a molecular picture of cononsolvency. *Macromolecules* 52(2): 457–64. https://doi.org/10.1021/acs.macromol.8b02196.

Zweep, N., A. Hopkinson, A. Meetsma, W.R. Browne, B.L. Feringa and J.H. van Esch. 2009. Balancing hydrogen bonding and van der waals interactions in cyclohexane-based bisamide and bisurea organogelators. *Langmuir* 25(15): 8802–9. https://doi.org/10.1021/la9004714.

Index